IN VIVO

The Cultural Mediations of Biomedical Science

PHILLIP THURTLE and ROBERT MITCHELL, Series Editors

IN VIVO

The Cultural Mediations of Biomedical Science

is dedicated to the interdisciplinary study of the medical and life sciences, with a focus on the scientific and cultural practices used to process data, model knowledge, and communicate about biomedical science. Through historical, artistic, media, social, and literary analysis, books in the series seek to understand and explain the key conceptual issues that animate and inform biomedical developments.

Life as Surplus

Biotechnology and Capitalism in the Neoliberal Era

MELINDA COOPER

A McLELLAN BOOK

UNIVERSITY OF WASHINGTON PRESS ❐ *Seattle and London*

This book is published with the assistance of a grant from the McLellan Endowed Series Fund, established through the generosity of Martha McCleary McLellan and Mary McLellan Williams.

Printed in the United States of America
Designed by Veronica Seyd
13 12 11 10 09 08   5 4 3 2 1

University of Washington Press
PO Box 50096, Seattle, WA 98145
www.washington.edu/uwpress

Library of Congress Cataloging-in-Publication Data

Cooper, Melinda.
Life as surplus : biotechnology and capitalism in the neoliberal era / Melinda Cooper.
p. ;   cm. — (In vivo)
Includes bibliographical references and index.
ISBN 978-0-295-98791-0 (pbk. : alk. paper)
1. Biotechnology—Political aspects—United States.   2. Life sciences—Political aspects—United States.   3. Capitalism—Health aspects—United States.   I. Title.
II. Series: In Vivo (Seattle, Wash.)
[DNLM: 1. Biological Sciences. 2. Capitalism. 3. Politics. QH 705 c777 2007]
TP248.2.c674   2007   303.48′3—dc22   2007038562

The paper used in this publication is acid-free and 90 percent recycled from at least 50 percent post-consumer waste. It meets the minimum requirements of American National Standard for Information Sciences—Permanence of Paper for Printed Library Materials, ANSI z39.48–1984.z

For Lucette,
Bob
and Melissa

Contents

Acknowledgments

M ANY PEOPLE COLLABORATED, WILLINGLY OR NOT, IN THE COMING TO LIFE of this book. Lucette has seen it through from its first inchoate stages to the final touches at Twigg's café in San Diego. This book owes so much to her and to her unique, adorable presence.

I would like to thank Françoise Duroux, for being such a generous, provocative doctoral supervisor; Brian Salter, for being a mover and shaker and for providing me with such an enabling environment at the University of East Anglia; Brian Massumi, for being so inspiring; Mick Dillon, Elspeth Probyn, Rosi Braidotti, and Joseph Dumit, for their professional and intellectual generosity; Kaushik Sunder Rajan, for being inimitable. I especially want to thank the friends who made Sydney such an intellectually exciting place to be—Anna Munster, Brett Neilson, Michael Goddard, and Jeremy Walker. I owe more than I can say to Catherine Waldby, friend and collaborator, who made this book and so much more possible. A thank you to Peter, for sporadic friendship and girl talk. And to Ingrid Renard, for being special.

I don't know how to thank Melissa for being so lovely and disarming, from Marrickville to Cairo.

I would like to thank my mother, Marina, and my sisters Wendy and Katina for their love and support; and my father, Bob Cooper, who died too soon, for being so sweet, encouraging, and politically inspiring.

Finally, I would like to thank the reviewers of the manuscript, Kaushik Sunder Rajan and Catherine Waldby (again). In the postmanuscript haze their incisive comments helped me see the light again, resulting, I feel, in a much clearer final draft.

This book was commissioned by Phillip Thurtle and Rob Mitchell, editors of the In Vivo series, and I am deeply grateful to them for being so encouraging. It has been a pleasure to work with Jacquie Ettinger, editor at the University of Washington Press. All of these people helped to make the daunting book-writing process not so bad at all.

LIFE AS SURPLUS

INTRODUCTION

T HE EARLY 1980S INAUGURATED AN ERA OF INTENSE CONCEPTUAL, institutional, and technological creativity in the life sciences and its allied disciplines. Not only did discoveries in molecular biology, cell biology, and microbiology promise to deliver new technological possibilities, they also called into question many of the founding assumptions of the twentieth-century life sciences. This was also the era of the "neoliberal revolution," where similarly dramatic transformations unfolded in the political, social, and economic spheres.[1] Initiated in the United Kingdom and United States, the neoliberal experiment sought to undermine the existing foundations of economic growth, productivity, and value, while at the same time it forged an ever-tighter alliance between state-funded research, the market in new technologies, and financial capital. In the United States in particular these interventions had a resounding effect on the life sciences. When President Ronald Reagan implemented a series of reforms designed to mobilize a "revolution" in the life sciences, public health, and biomedicine, he triggered a momentum that has been pursued by every administration since then. The project of U.S. neoliberalism, I argue throughout this book, is crucially concerned with the emergent possibilities of the life sciences and related disciplines.

As the realms of biological (re)production and capital accumulation move closer together, it is becoming difficult to think about the life sciences without invoking the traditional concepts of political economy—production, value, growth, crisis, resistance, and revolution. At the same time, however, the expansion of commercial processes into the sphere of "life itself" has a troubling effect on the self-evidence of traditional economic categories, compelling us to rethink their scope in dialogue with the life sciences themselves. The biotech era poses challenging questions about the interrelationship between economic and biological growth, resurrecting in often unexpected ways the questions that accompanied the birth of modern political economy. Where does (re)production end and technical invention begin, when life is put to work at the microbiological or cellular level? What is at stake in the extension of property law to cover everything from the molecular elements of life (biological patents) to the biospheric accident (catastrophe bonds)? What is the relationship between new theories of biological growth, complexity, and evolution and recent neoliberal theories of accumulation? And how is it possible to counter these new dogmatisms without falling into the trap of a neofundamentalist politics of life (the right-to-life movement or ecological survivalism, for example)?

Now, more than ever before, we need to be responsive to the intense traffic between the biological and the economic spheres, without reducing the one to the other or immobilizing one for the sake of the other. Working between the history and philosophy of science, science and technology studies, theoretical biology, and political economy, this book attempts to be as promiscuous in its investigations as the contemporary life sciences themselves.

Taking the North American commercial life sciences as its point of departure, this book does not presume to cover the entire history of contemporary biotechnology or to account for the widely different ways in which new life science technologies have been deployed and regulated around the world. As recent studies in the field have made abundantly clear, even the differences between such economic competitors as the United States, Britain, and Germany are stark enough to warrant careful comparative analysis (Jasanoff 2005). And the politics of a country such as India, with its unique history of drug production and patent.laws, brings an end to the idea that the emerging bioeconomies of the twenty-first century will be organized around rigid dividing lines between imperialist winners and postcolonialist losers (Sunder Rajan 2006). With the rise of East Asia as a significant hub of research and investment in the new life sciences, the global power dynamics of biocapital are far from determined in advance.

However, I contend that there is a specificity to the development of life science production in North America that demands analysis in its own right. This specificity lies as much in the recent history of U.S. economic crisis as in its present position as a focal point of world economic and imperialist power. As I show in chapter 1, 1980 marked a turning point in U.S. research and development (R & D) policy. Since then, the life sciences have played a commanding role in America's strategies of economic and imperialist self-reinvention. Over the past few decades the U.S. government has been at the center of efforts to reorganize global trade rules and intellectual property laws along lines that would favor its own drug, agribusiness, and biotech industries. Moreover, the unique position of the United States itself in relationship to world financial flows has meant that even the most speculative of its life science enterprises has attracted a constant, and incomparable, flow of funds. My perspective on imperialism therefore differs from that of post-autonomist Marxists Antonio Negri and Michael Hardt (2001), who argue that today there are no national focal points to world power relationships. On the contrary, I argue that U.S. nationhood occupies a central, if precarious position, in the constitution of global debt. This position is inseparable from America's engagement in the new life sciences.

POLITICAL ECONOMY AND BIOLOGY: GENEALOGIES

It is impossible to talk about biopolitics today without evoking a whole battle-ground of theoretical positions and counterpositions. As much as possible, I have avoided turning this book into a sustained dialogue with the existing theoretical literature on biopolitics, simply because I feel that many pressing questions are yet to be posed. However, the questions I formulate are prompted by two crucial moments in the work of the political philosopher Michel Foucault. The first of these can be found in Foucault's *Order of Things* (1973), in which he argues that the development of the modern life sciences and classical political economy should be understood as parallel and mutually constitutive events. Foucault locates the decisive turning point at the threshold of the late eighteenth and early nineteenth century, when the classical sciences of wealth (from the mercantilists to the physiocrats) were replaced by the modern science of political economy (Adam Smith and David Ricardo) and the natural history of the classical period (Comte Georges-Louis Leclerc de Buffon and Carolus Linnaeus) gave way to the science of life itself, the modern biology of Xavier Bichat and Baron Georges Cuvier. Prior to this, Foucault argues, there was no "life" in the modern, biological sense of the term, nor was there any conception of "labor" as the fundamental productive force underlying the exchange of money.

It is true, Foucault concedes, that the taxonomists of the classical period divided nature into the three classes of the mineral, vegetable, and animal, following Aristotle; at the same time, however, they accorded no particular salience to the distinction between the organic and the inorganic. If "life" existed as a category of classification, it could be made to slide from one end of the scale to the other, but it never constituted a self-evident threshold beyond which new forms of knowledge, science, and experiment would be necessary (ibid., 160). It is in this respect, according to Foucault, that the modern life sciences, whether vitalist or mechanist, represent a radical departure from the classical sciences of nature. At some point between 1775 and 1800 the opposition between organic and inorganic began to be perceived as fundamental, superimposing itself on the old order of three kingdoms and entirely reworking its categories of resemblance and difference (ibid., 232). From Bichat to Cuvier the conditions for a modern biology are established when life "assumes its autonomy in relation to the concepts of classification" and retreats from the order of visible relations into the physiological and metabolic depths of the organism (ibid., 162).

Foucault sees a similar transition at work in the founding texts of modern political economy, where the notion of labor as the indispensable and originary source of all value is articulated for the first time. It is in the work of Ricardo (rather than Adam Smith) that Foucault identifies the fullest realization of this transition: whereas the economists of the classical period saw value as a function of trade, exchange, and circulation, whose movements could be charted in the construction of elaborate economic tables, Ricardo inaugurated the properly modern science of economics by separating value from "its representativity" and relocating the source of all wealth behind the surface effects of exchange, in the time-processes of force, labor, and fatigue (ibid., 254). In the work of Ricardo, value for the first time ceased to be a mere sign of equivalence, circulating in the flat world of representation, and came to measure and be measured by something other than itself: the expenditure of force in time, "the human being who spends, wears out and wastes his life" (ibid., 257).

From Ricardo to Marx the science of economics now discovers production as the ultimate source of all value, whatever distortions it may later undergo in the sphere of exchange. And what this passage from representation to production reflects more generally, according to Foucault, is the relocation of wealth in the creative forces of human biological life rather than the fruits of the land—still evident, for example, in the work of Adam Smith. Herein lies the first point of articulation between economics and the modern life sciences: in the concept of "organic structure," Foucault writes, modern biologists discover a principle that "corresponds to labor in the economic sphere" (ibid., 227). In the nineteenth century the economy begins to grow for the first time, just as life comes to be understood as a process of evolution and ontogenetic development: the "organic becomes the living and the living is that which *produces, grows and reproduces*; the inorganic is the non-living, that which neither develops nor reproduces; it lies at the frontiers of life, the inert, the unfruitful—death" (ibid., 232).[2] As both Malthus and Marx make clear in their different ways, the question of population growth thereby becomes inseparable from that of economic growth. Henceforth, political economy will analyze the processes of labor and of production in tandem with those of human, biological reproduction—and sex and race, as the limiting conditions of reproduction, will lie at the heart of biopolitical strategies of power.[3]

BIOPOLITICS: FROM THE NEW DEAL TO NEOLIBERALISM

A second and more immediate point of reference for my argument is found in Foucault's seminar work on the rise of state biopolitics. In his 1975–76 lectures Foucault (2003) looks at the various strategies invented by the state in the course of the eighteenth and nineteenth centuries as a means of organizing the temporal processes of reproduction, disease, and mortality. Such strategies, he argues, are inseparable from the development of the mathematics of risk and statistical normalization—the bell curve or normal distribution as a way of standardizing and controlling the advent of future contingency on a collective level. Ultimately, what Foucault is pursuing here is the genealogy of the mid-twentieth-century welfare or social state—the constitutional form that most successfully brings together the administration of demographics with that of economic growth.

This aspect of Foucault's work has been explored most fully by the historian François Ewald, whose 1986 study *L'etat providence*, argues that the welfare state or New Deal social state is the first political form to place the actuarial strategies of risk socialization at the very core of government. The welfare state thus borrows its juridical forms from *life insurance*, generalizing its principles of mutual risk exchange to the whole nation. Unlike its liberal precursors the welfare state promises to take in charge the entire chronology of human life, from beginning to end. It is interested not only in the productive life of the laborer, but the reproductive life of the nation as a whole. The contract it establishes is one of mutual obligation, a mutualization of the biological in the service of the collective life of the nation. In this way, observes Ewald (1986, 326), the welfare state contract institutionalizes a form of collective, social generation: "The relationship of the individual to society is one of generation, kinship, inheritance," such that the protection of life becomes a political problem on a par with the reproduction of the nation, its continuity into the future.[4]

The welfare state is thus the first political form not only to understand obligation in immediately social, collectivist terms, but also to inscribe its relations of debt at the level of the biological. It undertakes to protect life by redistributing the fruits of national wealth to all its citizens, even those who cannot work, but in exchange it imposes a reciprocal obligation: its contractors must in turn give their life to the nation. In the philosophy of early welfare state advocates such as William Beveridge and Franklin Roosevelt, the economic survival of the nation is necessarily founded on this subterranean economy of biological reproduction: "The welfare state consolidates around the idea of the protection of the living. If the economy is the center of its preoccupations, as for the liberal state, this is not an economy of material wealth but an economy of life" (ibid., 375). And for Ewald (ibid., 327) the mid-twentieth-century invention of human rights discourse—with its appeal to a fundamental right to life—is merely the idealized expression of the welfare state economy of life: "The notion of a right to life is nothing but a principle of generalized socialization of existences, souls and bodies, a way of constituting them as infinitely indebted to society. . . . The notion of a right to life or right to existence is linked to an economy of obligations which is very different from the liberal economy. It demands to be formulated more in the form of duties than of rights. Society gives life and pledges to protect it. What does it ask for in exchange? That one gives one's life to society. . . . The counterpart to the right to life can only be the engagement, without reserve, of one's own life. The basis of the new language of rights is devotion."

This book is not primarily concerned with welfare state biopolitics and its colonial or developmental avatars.[5] Rather, I am interested in delineating the specific strategies of neoliberal biopolitics, as pursued by the United States on a domestic and global front over the past three decades. I therefore begin my analysis at precisely the point where Foucault left off in his 1978–79 lectures on *La naissance de la biopolitique* (2004), the first and last context in which he turned his attention to the rise of neoliberalism. In accord with Foucault, I start from the premise that neoliberalism reworks the value of life as established in the welfare state and New Deal model of social reproduction. Its difference lies in its intent to efface the boundaries between the spheres of production and reproduction, labor and life, the market and living tissues—the very boundaries that were constitutive of welfare state biopolitics and human rights discourse.[6]

As the British sociologist Richard Titmuss (1971) had predicted, the case of blood is exemplary here. When human blood is no longer treated as a national reserve and sequestered from the fluctuations of the market, the whole space of reproduction thereby becomes potentially available for commodification. Accordingly, what neoliberalism wants to capitalize is not simply the public sphere and its institutions, but more pertinently the life of the nation, social and biological reproduction as a national reserve and foundational value of the welfare state. In so doing, it undoes the constitutive mediations of the Keynesian social state, exposing the realm of reproduction to the harsh light of direct economic calculus.

At this point, however, my analysis of neoliberal biopolitics departs from that of Foucault. The *Naissance de la biopolitique* lectures focus on the theorists of the Chicago school of economics, for whom the neoclassical presumption of market equilibrium represents something like a law of nature. The neoclassical bias is reflected in the Chicago school theory of "human capital" and forms the basis of Foucault's critique of neoliberalism (2004, 221–44, 249). My study, however, accords more importance to those currents in neoliberal theory that have developed their own critique of neoclassical equilibrium models, currents that have been more closely associated with the rise of chaos and complexity theory, the return to Joseph Schumpeter's evolutionary economics (1934), and the later, complex models of self-organization proposed by Friedrich von Hayek (1969). In chapter 1, for example, I am concerned with the critique of rational expectations and equilibrium models of the market developed in the context of the Santa Fe Institute's conferences on economic theory. It is these nonequilibrium approaches, I suggest, that have exercised the greatest influence on the political and social forms of neoliberalism. And it is these theories, however deluded, that are most attuned to the actual conditions of labor and capital accumulation in the neoliberal era.

There is thus a preliminary distinction to be made between Keynesian and neoliberal understandings of growth.[7] Where social state growth strategies require the establishment of an institutional reserve or foundational value, neoliberalism divests itself of all national foundation, projecting its accumulation strategies into a speculative future. Essential to Keynesian economic strategies is the idea that the growth cycles of production, reproduction, and capital accumulation can be sufficiently calibrated to avoid capital's perennial catastrophe risks—labor insurgency and financial crisis (to which I add feminism and queer politics as a refusal of normative reproduction). Social state economics is a science of mediated growth, one that establishes institutional measures and foundational values from the reserve bank to fixed exchange rates and the family wage, as a means of warding off both social disruption and financial bubbles. Where welfare state biopolitics speaks the language of Gaussian curves and normalizable risk, neoliberal theories of economic growth are more likely to be interested in the concepts of the non-normalizable accident and the fractal curve. Where Keynesian economics attempts to safeguard the productive economy against the fluctuations of financial capital, neoliberalism installs speculation at the very core of production. Neoliberalism, in other words, profoundly reconfigures the relationship between debt and life, as institutionalized in the mid-twentieth-century welfare state. It does so in productive dialogue with the life sciences, where notions of biological generation are being similarly pushed to the limit.

I therefore propose a number of qualifications to Foucault's critique of neoliberalism. First, the operative emotions of neoliberalism are neither interest nor rational expectations, but rather the essentially speculative but nonetheless productive movements of collective belief, faith, and apprehension. What neoliberalism seeks to impose is not so much the generalized commodification of daily life—the reduction of the extraeconomic to the demands of exchange value—as its financialization. Its imperative is not so much the measurement of biological time as its incorporation into the nonmeasurable, achronological temporality of financial capital accumulation.[8]

In this sense too, neoliberal biopolitics returns us to forms of collective risk evaluation that were much more apparent in the nineteenth century, in the era prior to the consolidation of the actuarial, social state. In her 1979 book *Morals and Markets*, Viviana A. Zelizer provides an extraordinary inventory of the highly speculative forms of life "insurance" that proliferated in late-eighteenth- and early-nineteenth-century Europe and North America. In a context where the difference between speculation and risk hedging was far from evident, insurance policies on the lives of the poor and elderly were considered a legitimate form of investment, while popular lotteries regularly wagered on the chances of survival of the shipwrecked and newly arrived immigrants. Today, such practices can only elicit a shock of recognition. With its vested interest in biological catastrophism, neoliberalism is similarly intent on profiting from the "unregulated" distribution of life chances, however extreme. The difference, paradoxically, is that neoliberalism's catastrophism is much more organized. Moreover, it is much more materially capable than its classical liberal forebears.

The ground covered in this book spans the era of contemporary biotechnology, from the development of recombinant DNA to more recent interest in cell-based therapies, regenerative medicine, and stem cell science. While the first three chapters focus on recombinant DNA, molecular biology, and microbiology, the last three investigate the emerging fields of stem cell science and tissue engineering. Having said this, the problematics explored are not easily confined within individual life science disciplines, and I am more concerned with presenting a panorama of life science politics than providing the definitive history of any one life science technology.

Chapter 1 argues that the "biotech revolution" of the Reagan era needs to be understood in the larger context of the "neoliberal revolution" and its attempts to restructure the U.S. economy along postindustrial lines. In particular, it explores the crossover between neoliberal theories of growth, crisis, and limits and the strategies of speculative growth deployed in the development of new life science technologies. Neoliberalism and the biotech industry share a common ambition to overcome the ecological and economic limits to growth associated with the end of industrial production, through a speculative reinvention of the future. At the height of the high-tech euphoria of the 1990s, the biotech industry promised to overcome hunger, pollution, the loss of biodiversity, and waste in general, while the ecological and biopolitical problems associated with industrial capitalism only continued to worsen. This chapter casts a critical light on the rhetoric of perpetual growth associated with the biotech revolution, arguing that neoliberalism does not so much overcome industrial waste as displace it elsewhere—in space and time. It examines specific examples of proposed biotechnological solutions to the problems of industrialism (oil-eating bacteria, pesticide- and herbicide-resistant crops, bioremediation) as well as more extreme projects such as the attempt to re-create life on Mars. I develop the concept of "delirium" as a way of understanding the biotechnological project of reinventing life beyond the limit. This delirium, I argue, is inseparable from the dynamics of contemporary debt imperialism and the role of the United States within it.

In chapter 2, I narrow my focus onto the question of human surplus, the HIV/AIDS epidemic, and the structural violences of the contemporary pharmaceutical industry. Here I approach the problematic of debt imperialism from the point of view of sub-Saharan Africa, where the World Trade Organization's new patent laws and the pricing strategies of the U.S. and European drug conglomerates, in tandem with the rigors of debt servitude, have had devastating effects on the life chances of whole populations. The unfolding of the HIV/AIDS epidemic, I argue, is symptomatic of the biopolitical form of contemporary debt imperialism. What it brings to light is the imperative of violence that sustains neoliberalism's promise of more and better life. I locate the rhetorical form of neoliberal violence in contemporary discourses of biological security and the humanitarian concept of "complex emergency."

However, I do not mean to establish a simple oppositional relationship between the imperialist center and its peripheries, or to suggest that U.S. debt imperialism is solely outward looking, turned toward the generic "non-West." On the contrary, it seems to me that the project of IMF- and World Bank–sponsored neoliberal imperialism—the so-called Washington Consensus—can only be properly understood as a strategy directed simultaneously against the United States' own underclasses and those of the developing world. Nor do I wish to deflect from the equally culpable politics of the post-apartheid Mbeki government. President Thabo Mbeki's politics, I argue, are symptomatic of the dangers of a neonationalist response to global imperialism.

Chapter 3 is concerned with the "biological turn" in the war on terror. Under President George W. Bush it has become increasingly difficult to distinguish public health policy on emerging infectious disease from U.S. military doctrine on so-called emerging threats, while the future of life science research is being rerouted toward military applications. What is at issue here, it is argued, is the extension of the speculative logic of financial capital into the military sphere and life science research. Moreover, Bush-era policy on infectious disease and biological warfare amounts to a turning inward of the strategies of humanitarian warfare, so that the "complex emergency" is rediscovered on American soil. This chapter is particularly interested in the concepts of preemption, emergence, and the catastrophe risk as each has developed at the intersection of economics and the life sciences.

The second part of the book turns to an examination of the science and politics of stem cell science and regenerative medicine. Chapter 4 provides an in-depth analysis of current experiments in tissue engineering and compares its methods with those of such earlier twentieth-century body technologies as organ transplantation and prosthetics. I have chosen to begin with a consideration of the technologically driven field of tissue engineering, rather than the more theoretical developments associated with "basic" stem cell biology, because I feel that experiments in bodily transformation are themselves informing new concepts of cellular potentiality, plasticity, and malleability. Tissue engineering, I argue, implicitly draws on topological methods of transformation. In contrast to earlier twentieth-century organ technologies, it is more concerned with processes of permanent embryogenesis than states of suspended animation. Again, a complex interchange between biological and economic epistemologies is at work here, since post-Fordist modes of production are similarly attuned to the possibilities of topological transformation.

Having outlined the modes of bodily transformation at work in emerging post-Fordist technologies of the body, I then turn to the more specific question of bodily generation, as it has been reconfigured by stem cell science. Chapter 5 is therefore concerned with the interfaces between reproductive medicine and stem cell science. It attempts to delineate the different concepts and technologies of bodily generation at work in these two fields, and the ways in which they come together as part of an emerging reproductive economy. I am interested both in the distinctive forms of commercialization that have accompanied the rise of stem cell science and the shifting economies of feminized, reproductive labor that it has already called into being. I argue that the delirium of contemporary capital finds its most extreme manifestation in the self-regenerative, embryoid bodies of stem cell science, and therefore calls for a new critique of the (reproductive) political economy.

Chapter 6 turns to the most conservative, fundamentalist impulses at work in contemporary neoliberal biopolitics. Here I am concerned with the rise of the American evangelical right and its culture of life politics. What are the connections between evangelical doctrines of personal rebirth, faith, and capital, and the politics of the contemporary right-to-life movement? Moreover, how can we understand the complex articulations between a neoliberal and neofundamentalist politics of life? Here again, I return to the problematic of debt imperialism and U.S. nationhood, which I examined in chapter 1, arguing that evangelical, right-to-life politics is inseparable from the promise and violence of American indebtedness and its role in the world economy today.

Each of these chapters seeks to illuminate a particular facet of life science politics in the neoliberal era. Some of these chapters are primarily concerned with the promissory, transformative, and therapeutic dimension of the contemporary life sciences. In other chapters I am more interested in exploring the forms of violence, obligation, and debt servitude that seem to be crystallizing around the emerging bioeconomy. However, I do not mean to suggest any linearity or finality to this sequence of ideas. Perhaps the starkest contrast is between the third and fourth chapters, where I move from the militarization of life science research to the regenerative possibilities opened up by stem cell science and tissue engineering. But even here the apparent contrast calls for qualification, as stem cell science is also being developed toward military ends (notably with the production of cell biosensors), while it is highly possible that the money being poured into biodefense research will yield unexpected therapeutic discoveries. In my organization of chapters and juxtaposition of ideas I hope to convey a sense of my own indecision as to the biopolitical futures enabled by contemporary life science production. As so much of contemporary biology insists, these futures can never be determined in advance.

□ 1 □

LIFE BEYOND THE LIMITS

Inventing the Bioeconomy

> Normal is passé, extreme is chic. While Aristotle cautioned "everything in mod-
> eration," the Romans, known for their excesses, coined the word "extremus," the
> superlative of exter (being on the outside). By the fifteenth century "extreme"
> had arrived, via Middle French, to English. At the dawning of the twenty-first
> century we know that the Solar System, and even Earth, contain environmen-
> tal extremes unimaginable to the "ancients" of the nineteenth century. Equally
> marvellous is the detection of organisms that thrive in extreme environments.
> [Biology has] named these lovers of extreme environments "extremophiles."
> —Lynn J. Rothschild and Rocco L. Mancinelli, "Life in Extreme Environments"

THE CONTEMPORARY BIOTECH INDUSTRY WAS BORN IN A CONTEXT
of intense speculation about the future of U.S. science and technology. After
acting as the motor of international economic growth in the decades following World War II, the United States was traversing a period of decline, whose
effects on world economic relations had yet to become clear. This period—which
can be situated roughly between the late 1960s and mid-1970s—saw an
extraordinary outpouring of futurological literature attempting to divine the
economic and political futures open to the United States and its competitors.
It also gave rise to the genre of global forecasting, in which the future of the
earth itself was put to the test of computer-based systems analysis.[1]

Undoubtedly, one of the most powerful documents of crisis produced in this
period was the Club of Rome's world futures report of 1972 (Meadows et al.
1972). Under the direction of a team of systems analysts based at Massachu-
setts Institute of Technology (the so-called Meadows team), the report gave voice
to the prevailing consensus that Fordist manufacture had entered a period of
irreversible decline. But it also brought something palpably new to the analy-

sis. If there was a crisis in the offing, it was not one that could be measured in conventional economic terms—a crisis in productivity or economic growth rates—but rather a wholesale crisis in the realm of reproduction. For the Club of Rome what was at stake was no less than the continuing reproduction of the earth's biosphere and hence the future of life on earth. The most visible signs of the impending crisis were therefore to be found in the existence of all kind of ecological disequilibria, exhaustion, and breakdown, from rising levels of pollution to famine and the increase in extinction rates. Using the latest developments in systems theory, the Meadows team sought to simulate the earth's possible futures by looking at trends and interactions between five principle areas—population growth, industrialization, food production, the depletion of nonrenewable resources, and pollution.

From the beginning, the report stressed the impossibility of arriving at any precise predictions, and yet in spite of variations, repeated runs of the simulation program pointed to one constant: the exponential growth of population and industry could not continue indefinitely without running up against the limits inherent in the other variables under study—namely, agricultural production, energy supplies, and pollution. Pointing out that 97 percent of industrial production, including agriculture, was dependent on such fossil fuels as natural gas, oil, and coal, the report anticipated that continued economic growth would soon come up against insurmountable limits. These limits were of two kinds, consisting not only in the depletion of nonrenewable resources but also in the steady environmental buildup of toxic, nonbiodegradable wastes. In other words, for the Club of Rome economic growth was synonymous with industrial production and would therefore end up faltering before the earth's geochemical limits. Already in the early 1970s there were signs that increasing levels of carbon dioxide in the atmosphere—"thermal pollution" as it was then called—could seriously disrupt the earth's climate and ecosystems (Meadows et al. 1972, 73). Such portents were all witness to a "simple fact," according to the Meadows team—"the earth is finite"—and even though we can't calculate the upper limits to growth with any precision, there are definite limits nevertheless (ibid., 86).

Twenty years later, and armed with more sophisticated modelling tools, the same team came up with a slightly more nuanced prognosis for the future. Limits to growth, they now argued, were time-like rather than space-like. This meant that we might have already gone beyond the threshold at which an essential resource such as oil could be sustainably consumed, long before we would notice its actual depletion. In fact, it was highly probable, according to the report's authors, that we were already living beyond the limit, in a state of suspended crisis, innocently waiting for the future to boomerang back in our faces: *"Time* is in fact the ultimate limit in the World3 model—and, we believe, in the 'real world.' The reason that growth, and especially exponential growth, is so insidious is that it shortens the time for effective action. It loads stress on a system faster and faster, until coping mechanisms that have been able to deal with slower rates of change finally begin to fail" (Meadows et al. 1992, 180). The conclusions of the 1992 report, however, remained substantially the same: in order to sustain life on earth, economic growth would need to respect ecological and biological equilibria. The current tendency toward exponential growth would need to be replaced by a steady-state economy.

The political consequences of the Club of Rome report were so resounding that President Jimmy Carter commissioned a follow-up document, *The Global 2000 Report*, drawing on the precise statistical data produced by various government departments and agencies in order to continue the work of prediction up until the year 2000—when life, it anticipated, would be even more precarious (Council of Environmental Quality and U.S. State Department 1980). The report was produced in a context of unprecedented government legislation around environmental issues—from the banning of pesticides to the passage of antipollution laws and the establishment of the Environmental Protection Agency (EPA).

But already in the 1970s the Club of Rome had set off a virulent counterreaction on the part of the new right. For Daniel Bell, one of the leading prophets of the postindustrial economy, the problem with the Club of Rome report was that it assumed what it set out to prove. Its very model of growth, based on a "simplified quantitative metric" and a "closed system," was bound to run up against limits sooner or later. It was incapable of thinking through the kind of systemic "qualitative change" that, according to Bell (1974, 464), characterized the successive phases in the evolution of capital. And for Bell, precisely such a change was required in the move from an industrial to a postindustrial economy. Throughout the 1970s theorists of the new right called for a radical restructuring of the U.S. economy. In order to reassert its world dominance, it was claimed, the United States would need to move from heavy industry to an innovation-based economy, one in which the creativity of the human mind (a resource without limits) would replace the mass-production of tangible commodities.

The postindustrial literature, however, was never concerned solely with the immaterial, innovation-based aspects of economic growth; one aspect of this literature that has been consistently overlooked is its claim to have found a solution not only to economic decline but also to environmental crisis. According to right-wing futurologists employed especially for the purpose by President Reagan, the postindustrial economy would not only take economic growth beyond the limit, it would also respond point by point to the ecological and biospheric limits painstakingly detailed by the Club of Rome (Simon and Kahn 1984). In particular, these theorists pointed to the promises of biotechnology as a way of internalizing, and thus overcoming, all limits to growth—from the waste products of industrialism to the very finitude of the earth: "Each epoch has seen a shift in the bounds of the relevant resource system. Each time, the old ideas about 'limits,' and the calculation of 'finite resources' within the bounds, were thereby falsified. Now we have begun to explore the sea, which contains amounts of metallic and perhaps energy resources that dwarf the deposits we know about on land. And we have begun to explore the moon. Why shouldn't the boundaries of the system from which we derive resources continue to expand in such directions, just as they have expanded in the past?" (Simon 1996, 66). In response to the Club of Rome's quasi-cosmological warnings that the future was nearing exhaustion, the postindustrialists cited the work of physicist Freeman Dyson to affirm that time was without limit (ibid., 65). And if cosmological time was without limit, then time, become immanent to matter, would regenerate the earth.

With its promise of future surplus on earth and beyond, the postindustrial literature set the scene for Reagan-era science policy—a policy that combined virulent antienvironmentalism and cutbacks in redistributive public health with massive federal investment in the new life science technologies and their commercialization. But what began as a utopian polemic designed to justify the specific machinations of the Reagan administration has since become the mainstay of neoliberal orthodoxy and as such has travelled far beyond the ranks of committed Reaganites and new right think tanks. Under President Clinton in particular, and during the stock market boom years of the late 1990s, the neoliberal promise came to be associated with a kind of libertarian, free-market vitalism. It was during this period that the concept of the "bioeconomy" began to take shape, culminating in the Organisation for Economic Cooperation and Development's decision to launch a major policy project in the area (OECD 2004).

The aim of this chapter is to provide a genealogy of the ideas and institutions that have brought the promise of the bioeconomy into being. My premise, as I have explained in the introduction, is that the emergence of the biotech industry is inseparable from the rise of neoliberalism as the dominant political philosophy of our time. The history of neoliberal theories of economic growth and biotechnological visions of growth therefore needs to be pursued simultaneously. I am interested in the specific conjuncture of domestic economic crisis that afflicted the United States in the early 1970s and the role of the life science industries in promoting a certain response to that crisis. The biotech revolution, I argue, is the result of a whole series of legislative and regulatory measures designed to relocate economic production at the genetic, microbial, and cellular level, so that life becomes, literally, annexed within capitalist processes of accumulation. Part of my work here is to detail the specific forms of property right, regulatory strategies, and investment models that have made this possible.

These questions lead me later in the chapter to a consideration of the important shifts in world imperialist relations that have occurred since the late 1970s (and in particular since the monetarist counterrevolution of 1979 through 1982). This period, according to political economists such as Giovanni Arrighi and Michael Hudson, has been one in which nation-state imperialism and the role of the United States within it have undergone a series of dramatic transformations (Arrighi 2003; Hudson 2003 and 2005). Here, I go further and interrogate the links between this shift in world imperialist relations and the growing importance of biological life within capitalist accumulation strategies. My argument here, and throughout the book, is that the geopolitics of world imperialism, as established in the post–World War II era, is today being displaced by a new and relatively mutable set of biopolitical relations, whose dynamics have yet to be theorized in detail.

In this context a number of methodological and conceptual questions become imperative. When capital mobilizes the biological, how do we theorize the relationship between the creation of money (surplus from debt; futures from promise) and the technological re-creation of life? Has the one been co-opted into the other? When capitalism confronts the geochemical limits of the earth, where does it move? What is the space-time—the world—of late capitalism, and where are its boundaries? What finally becomes of the critique of political economy in an era in which biological, economic, and ecological futures are so intimately entwined? And when the future itself is subject to all kinds of speculation?

This aspect of my work returns to Karl Marx's still fertile reflections on crisis, limits, and growth, in order to discern what is peculiar to the neoliberal moment in capital accumulation. My starting point is, in one sense, classically Marxian: I take for granted that the periodic re-creation of the capitalist world is always and necessarily accompanied by the reimposition of capitalist limits; that capitalist promise is counterbalanced by willful deprivation, its plenitude of possible futures counteractualized as an impoverished, devastated present, always poised on the verge of depletion. Yet my analysis extends into a sphere to which Marx paid relatively little attention: that of the life sciences conceived in the broadest sense of the term. For this reason I am also concerned with contemporary theoretical and technical developments in biology, the environmental sciences, and evolutionary theory. Recent biology, I argue, is as much interested in the limits and possible futures of life on earth as contemporary capitalism. Any critique of the bioeconomy therefore needs to address itself to the intense traffic of ideas between recent theoretical biology and neoliberal rhetorics of economic growth.

What this critique requires, I suggest, is not so much an analysis of market fetishism, simulacra or phantasm, as of capitalist delirium. Freud tells us that the psychotic delirium, as opposed to the neurotic fantasy, is crucially concerned with the breakdown and recreation of whole worlds. Delirium is systemic, not representative. It seeks to refashion the world rather than interpret it. In this respect the concept of delirium has obvious affinities with Marx's reflections on the self-transformative, world-expansive tendencies of capital.[2] But delirium is no less evident in the rhetoric of the biorevolution, where speculative meditations on the future of life on earth are never far from the agenda. This rhetoric isn't merely peripheral to the real business of the life science industries. Rather, the delirium of contemporary capitalism, I argue, is intimately and essentially concerned with the limits of life on earth and the regeneration of living futures—beyond the limits.

In this sense the concerns voiced by the Club of Rome and its critics express, in their different ways, the conflicting tendencies that animate this delirium.

Most strikingly perhaps, the delirium finds expression in the NASA space biology program—a program whose conceptual and economic influence on the biotech revolution has been curiously neglected. In the words of its director, the not so humble aim of the NASA space biology program is "'to improve life here . . . to extend life there . . . to find life beyond'" (cited in Dick and Strick 2004, 230). The program's influence, I suggest, is increasingly present in the more practical applications of life science research, including, most recently, those proposed by the OECD's 2005 report on the bioeconomy and the U.S. Energy Act of 2005. By moving to and fro between the cosmic futures opened up by space biology and the mundane world of industrial and commercial biotech policy, I hope to show how the delirium of late capitalism gets translated, in very pragmatic fashion, into the day-to-day infrastructures of government and science. The interest of this method is to develop a critique that is at once sensitive to the global, systematizing momentum of capitalist dynamics and the micropolitical decisions that bring it into being. This method also suggests ways in which the delirium of capital can be challenged on a practical level.

RESPONDING TO CRISIS: REGENERATING WASTE

The details of the American economic crisis of the late 1960s have been rehearsed elsewhere. Throughout the 1970s the United States went from stagflation to recession as two successive oil shocks drastically raised the production costs of the manufacturing industry. U.S. industries began to face declining profitability as Japanese and European exports became increasingly competitive in international and domestic markets. By the late 1970s the ailing fortunes of industry had been compounded by a looming financial crisis, as American multinationals increasingly chose to invest their surplus dollars in offshore money markets rather than repatriate them in the United States.[3] However, the decisive impact of this moment of crisis in shaping the future life science industries has been less closely explored.

The crisis had a particular impact on the whole arena of chemical production, which extended from plastics, fabrics, and such agricultural commodities as fertilizers and herbicides—the very stuff of Fordist mass manufacture and monoculture—to the pharmaceutical industry. It was largely at the initiative of these industries that molecular biotech would be born as a commercial venture. The U.S. petrochemical industry had flourished as a mass producer of tangible commodities during the 1950s and 1960s, but by the 1970s was faced with steeply declining profits, which were only exacerbated by the oil price shocks of 1973 and 1979.[4] Moreover, the profits that had come from exporting Green Revolution mass monoculture to the developing world were beginning to fall, at least partially as a result of its devastating environmental consequences.[5] Within the United States itself, mounting pressure from the green movement, combined with increasing governmental regulation, meant that chemical industries were being forced to internalize the costs of their own waste production.

For the pharmaceutical industry the period of decline came slightly later but was just as decisive in forcing it to reorient its commercial strategies.[6] Unlike petrochemicals, the fortunes of the pharma industry had long been based on patent protection and innovation monopolies. Microbiology and organic chemistry, protected by chemical patents, had fueled the post–World War II boom in drug innovation. By the late 1970s, however, generics were beginning to flood the market while the rate of drug innovation had declined dramatically. Again, this downturn can be ascribed at least in part to the effects of growing public concern surrounding the nonregulation of clinical trials (many of them conducted in prisons) and the toxicity of certain wonder drugs. In the wake of thalidomide and other prescription drug disasters, government regulation had become much tighter and the lengths of clinical trials significantly prolonged. It was thus in response to the commercial limits posed by regulation (rather than ecological limits) that the petrochemical and pharmaceutical sectors began to reorganize. As prescribed by Daniel Bell, their response was not to falter in the face of undeniable limits but rather to relocate beyond the limits of industrial production—in the new spaces opened up by molecular biology.

The 1980s was a period in which the U.S. petrochemical and pharmaceutical industries embarked on a dramatic self-imposed makeover, reinventing themselves—at least prospectively—as purveyors of the new, clean life science technologies. Thus by the early 1980s all of the major chemical and pharmaceutical companies had invested in the new genetic technologies, either through licensing agreements with biotech start-ups or by developing their own in-house research units.[7] Even for a notorious toxic-waste offender such as Monsanto, it now seemed clear that the extractive, petrochemical industries that had fueled the boom years of the Fordist economy were destined to be subsumed within the new paradigm of post-Fordist bioproduction. Taking the lead from recent successes in recombinant DNA and availing themselves of new patent laws, companies such as Monsanto began experimenting with all kinds of novel ways for reincorporating their former investments in petrochemical processes into the ambit of biomolecular science.

The commercial calculus was straightforward—instead of profits from mass-produced chemical fertilizers and herbicides, the agricultural business would displace its claims to invention onto the actual generation of the plant, transforming biological production into a means for creating surplus value. Moreover, it was predicted that biotechnology would expand the geological spaces open to commercial agriculture, making it possible to create plants that would survive on arid land or flourish in the degraded environments created by industrialized agriculture.[8] Indeed, according to some prognoses, life itself would soon be put to work to remediate all kinds of industrial waste—from chemical pollutants to nuclear fallout (so-called bioremediation). In short, the geochemical laws ruling over Fordist industrial production would be replaced by the much more benign, regenerative possibilities of biomolecular production.

In this process of transformation, two tendencies have been at work. On the one hand, the pharmaceutical and petrochemical industries have responded to crisis by initiating an extraordinary internal consolidation of all aspects of the commercial life sciences, with the result that a handful of transnational (but all U.S.- and EU-based) companies now effectively control every level of world food and pharmaceutical production.[9] On the other hand, the same companies have preemptively moved to capture new and emerging markets in life science production by establishing strategic alliances with smaller biotech companies. The phenomenon of the biotech start-up has in turn been enabled by a systemic recourse to financial capital and financial modes of investment.

For the former producers of Fordist petrochemicals then, the imperative to self-reinvent signified a move not only into a new space of production—molecular biology had until then remained an area of basic science research—but also into a new regime of accumulation, one that relies on financial investment to a much greater extent than the Fordist economy had. The political economists Michel Aglietta and Régis Breton (2001) have argued that financial liberalization, in the United States at least, has inaugurated a new "finance dominated regime of accumulation" in which the evaluation of future profits becomes the decisive factor in determining price. Whether we accept the details of their argument, this has certainly been the case for the U.S. biotech sector, where a combination of venture capital funding and stock market initial public offerings (IPOs) have become the standard business model for emerging companies.[10] Here, it would seem, the financial markets have become the very generative condition of production, making it impossible to distinguish between the so-called economic fundamentals and the perils and promises of speculation. Indeed "promise," it might be argued, is the one fundamental of post-Fordist production: promise is what enables production to remain in a permanent state of self-transformation, arming it with a capacity to respond to the most unpredictable of circumstances, to anticipate and escape the possible "limit" to its growth long before it has even actualized.

As a consequence, biological production, like any other area of post-Fordist enterprise, has undergone a dramatic destandardization. Post-Fordism relies much more immediately on economies of innovation, scope, and preemption— the ability to anticipate the next wave, to keep ahead of the curve—than the economies of scale associated with the mass reproduction of commodities.[11] What counts here is the variable code source from which innumerable life forms can be generated, rather than the life form per se. Hence the biological patent allows one to own the organism's *principle of generation* without having to own the actual organism. In the age of postmechanical reproduction the point is no longer to reproduce the standardized Ford-T model in nature, but to generate and capture production itself, in all its emergent possibilities. Its success is dependent on the constant transformation of (re)production, the rapid emergence and obsolescence of new life forms, and the novel recombination of DNA rather than the mass monoculture of standardized germplasm.

This is not to suggest that in the recombinant era mass reproduction has been made obsolete; rather, it has been demoted as the principle source of surplus value and subsumed within a higher-order mode of production. Henceforth, profits will depend on the accumulation of biological futures rather than on the extraction of nonrenewable resources and the mass production of tangible commodities. Nor does the move into bioproduction signify the dissolution of ecological limits. The new life science conglomerates have not overcome waste, depletion, or any other of the catastrophic limits to life on earth, but they have simply divested themselves of the costs. Post-Fordism does not dispense with industrial production; it simply displaces it—either literally, by moving it offshore, or legally, by fighting for deregulation. In the process post-

Fordism allows itself to disregard the effects of waste production entirely. Post-Fordist modes of production may thus go hand in hand with an intensification of industrial pollution, as is visible in the tons of computer hardware cast off by the information economy and in the masses of chemical inputs that can be used with impunity on a genetically modified (GM) herbicide-resistant crop. By relocating in the promissory future of the life sciences, what post-Fordism has overcome, at least temporarily, is the decline in profits associated with petrochemical production.

The drive to overcome limits and relocate in the speculative future is the defining movement of capital, according to Marx. Yet there is one limit that capitalism never escapes—the imperative to derive profit and thus to recapture the "new" within the property form. For Marx capitalism is essentially counterproductive.[12] It cannot expand into a new space-time of accumulation, beyond its actual limits, without bringing this one internal limitation along with it. Thus, even when it seems to move into the most evanescent and unexpected of futures, it will need to subtract from the very surplus it calls into being. Only on this condition can the promise of a surplus life be rendered profitable. The willful production of waste is a capitalist imperative common to the industrial and postindustrial eras. The difference lies merely in their temporalities: while industrial production depletes the earth's reserves of past organic life (carbon-based fossil fuels), postindustrial bioproduction needs to depotentialize the future possibilities of life, even while it puts them to work. This counterlogic is perhaps most visible in the use of patented sterilization technologies, where a plant's capacity to reproduce itself is both mobilized as a source of labor and deliberately curtailed, thus ensuring that it no longer reproduces "for free." But it is also endemic to the whole enterprise of capitalized bioproduction.

RULES AND REGULATIONS: CREATING THE BIOTECH REVOLUTION

It would be tempting to see the rise of the U.S.–based bioeconomy as some kind of inevitable, even redemptive moment in the evolution of capital—much of the neoliberal, new economy literature does precisely this. But in an important sense the U.S. biotech industry was actively fostered, promoted, and brought into being from the top down by a series of quite deliberate legislative and institutional decisions.[13] In the mid-1970s new technological possibilities were undoubtedly emerging in the life sciences, particularly from molecular biology, but these were (and in many cases still are) far from the stage of advanced clinical or field trial and commercialization. In this sense the invention of the North American biotech industry needs to be understood as a speculative maneuver, but one that was underwritten by rigorous federal protections. Indeed, the very promissory nature of the biotech economy is itself the result of a whole range of reforms adopted in the Reagan era. These reforms have transformed the nature of life science research in such a way that the mere hope of a future biological product is enough to sustain investment.

This process reflects the response of successive administrations to an evolving rhetoric of crisis and transformation in the U.S. economy. The closing years of the Carter administration had produced an abundance of private and government-funded reports claiming to diagnose the reasons behind America's declining profits.[14] In general, these reports ascribed the downturn to two government failures in the area of science funding. First, it was claimed that in the absence of relevant intellectual property laws, what the United States had produced in terms of R & D was too easily replicated by such newly industrialized competitors as Japan. The North American Fordist model, it appeared, had been trumped by Japanese "post-Fordism." Inventions that had originated in America at the height of its industrial Golden Age could now be mass produced with greater efficiency and at lower cost by its rivals, who would then promptly send them back to flood the North American domestic market.

The advocates of reform went on to argue that federal funding of science placed undue emphasis on basic research at the expense of technical applications and commercial outcomes. What was called for was a fundamental reworking of the relations between academic science and the private sector, public research funding, and commercial interests, so that these groups could work together more closely. Moreover, a number of specific recommendations were made concerning the state's future role in relationship to science research: the government should restore public funding of R & D to its former high levels but should otherwise remain absent from decision making about research directions; at the same time the government should create incentives for private-academic alliances and encourage commercial enterprise to intervene in the final—and profit-making—stages of product development.

In response to these reports, the incoming Reagan administration in 1980 implemented a series of reforms that would set the scene for U.S.–based life science research over the following decades. The most immediate result of the reforms was a dramatic increase in federal support for the life sciences, which henceforth would become the most heavily funded area of basic science research in the United States, apart from defense. The United States now dedicates more of its federal budget to science research than any other OECD country. The National Institutes of Health (NIH), which funds a majority of life science research projects, absorbs up to 60 percent of this budget. Since Reagan, funding of the life sciences has only continued to increase. However, this focus on the life sciences has gone hand in hand with a redistribution of funds away from public health and nonprofit medical services toward commercially oriented research, health services, and for-profit applications.[15]

Among the many reforms in R & D, investment, and industrial policy carried out throughout the 1980s, a handful can be singled out for their decisive impact in shaping the future direction of the life sciences. In 1980 the Patent and Trademark Amendments (or the Bayh-Dole) Act was passed, setting the stage for the decidedly entrepreneurial, public-private alliances that would come to characterize the whole field of life science research over the next few decades. Henceforth, publicly funded science institutions and labs would not only be authorized but well-nigh obliged to patent the results of their research. These initially public-funded patents could then be privately exploited by the patent holders, who might choose to issue exclusive licences to large private companies, enter into joint ventures, or create their own start-up companies. The Bayh-Dole Act effectively gave rise to a new academic personage, the scientist-entrepreneur, and a new form of public-private alliance, the joint-venture start-up, in which academics and venture capitalists come together to commercialize the results of public research.

A crucial element in the success of these public-private life science alliances was the rise of essentially uncertain, high-risk forms of investment. In this respect the United States enjoys a considerable advantage over its competitors. The deregulation of banking and financial markets that took place throughout the 1970s, a highly liquid stock market, and the securitization of pensions have all meant that a large source of funds are available for investment in emerging high-risk ventures. And in an environment already conducive to speculative investment, the United States has moved to consolidate its financial edge.[16] In the course of the 1980s new legislation made it possible for large institutional investors such as pension funds, awash with the newly privatized savings of American workers, to invest a percentage of their portfolios into high-risk shares, junk bonds, and (most important for biotech) venture capital funds.

A second and equally important reform was the establishment of the NASDAQ as a market for venture securities. Unlike conventional exchanges the NASDAQ was able to list high-risk start-up firms that had registered losses for several years running and boasted little or no collateral. These unprofitable firms were authorized to include a whole range of intangible, speculative assets in their financial statements, including patent portfolios on not yet commercialized products. In this way the stock market business model has worked in conjunction with the evolution of patent law to institutionalize an essentially promissory market in life science innovation. In the absence of any tangible assets or actual profits, what the biotech start-up can offer is a proprietary claim over the future life forms it might give rise to, along with the profits that accrue from them. In essence then, what these reforms have formalized is the prospective value of promise, turning life science speculation into a highly profitable—indeed rational—enterprise.[17]

Again, it was the Reagan administration that made the first moves to reform existing intellectual property laws, with the long-term aim of incorporating all kinds of biological products and processes within the scope of patentable invention. Although patent laws on biological products have subsequently been established at various national and regional levels worldwide, the U.S. model of patent law remains by far the most liberal.[18] Importantly, these domestic reforms took place against the backdrop of efforts to neoliberalize conditions of international trade in such forums as the World Trade Organization. Hence, as legal theorist Susan K. Sell (2003) has shown, the same private interests who had been behind the push for liberalized patent law within the United States also embarked on a concerted campaign to impose these extremely favorable new laws internationally throughout the 1980s, culminating in 1986 with the signing of the trade-related intellectual property rights (TRIPs) agreement.

I discuss the TRIPs agreement in chapter 2. Here I want to simply note the special circumstances that enabled the United States to massively promote its life science industries at a time when its competitors were all cutting back on research funding. Various commentators have noted the singular position of the United States with respect to global financial flows over the past few decades. Whereas the investment decisions of most countries—in particular with respect to R & D—have come under the restrictive control of global capital, the United States alone has been able to pursue a program of profligate domestic expenditure, especially in the areas of defense and life science research. While most OECD countries reduced their levels of R & D funding in the 1990s, the United States increased them significantly—much of this increase going to the NIH, the principle funding body for biotech and pharmaceutical research.[19] Since the early 1980s, the continued ability of the United States to lavish money on life science research has been kept afloat by its unique position at the focal point of world financial flows and as the world's largest debtor.

WORLD ECONOMIES: ON DEBT CREATION, LIMITS, AND THE EARTH

In his account of the shifting basis of world imperialism over the past three decades, the economist Michael Hudson (2003 and 2005) has explained how the United States, since abandoning the gold standard, has transformed the U.S. Treasury bill, a government-issued debt, into an international monetary standard, a promissory note to U.S. power in which all other national banks are more or less obliged to invest. Prior to 1971, the oil-dollar-gold standard had functioned as a check to the U.S. government's ability to run balance-of-payments deficits without limit, since foreign central banks could always hold the world's banker to account by exchanging their surplus dollars for gold. When gold was demonetized, however, all institutional standards of measurement went with it. With foreign banks unable to exchange their dollars for gold, no other option was available to them than that of buying up U.S. Treasury bonds. In other words countries that held a surplus of dollars could do only one thing with them—purchase U.S. Treasury debt, placing them in the unenviable position of extending a continuous loan to the U.S. government. As it stands, writes Hudson (2003, xv), this loan has become so structural to world economic relations that it demands to be "rolled over indefinitely" and "will not have to be repaid."

These tendencies, Giovanni Arrighi has argued (2003, 62–66), were consolidated by the monetarist counterrevolution of 1979–82, when the United States introduced interest rate policies that had the overall effect of refunneling global financial flows back into the dollar and the U.S. markets. It is this move that accounts for the American economic revival of the 1990s—and its enormous R & D investments in the life sciences. Henceforth, North American imperial power would be reestablished on a new and paradoxical basis: having acted as the world's principle creditor nation in the years following World War II, the United States would now assert itself as the world's largest debtor, inflating its capital markets and fueling its spiraling budget deficits with a continuous inflow of foreign capital investment. In the process the United States has plummeted to a level of indebtedness that *"has no precedent in world history"* (ibid., 70). In this way the early 1970s set off a process by which the United States transformed itself into the focal point of an effective debt imperialism—a world empire that is curiously devoid of tangible reserves or collateral, an empire that sustains itself rather as the evanescent focal point of a perpetually renewed debt and whose interests lie in the continuous reproduction of the promise.[20]

What becomes of money when the debt form goes global? Indeed, what becomes of imperialism when the world's greatest power derives its funds from an influx of perpetually renewed debt? For Marx the creation of money from debt represents the most insane form of the capitalist delirium. It is here, he writes, that capital begins to imagine itself as self-valorizing value: a life force possessed of its own powers of self-regeneration.[21] Hudson has argued that contemporary U.S. debt imperialism takes this delirium to its logical extreme. The establishment of U.S. Treasury bills as an international monetary standard, he claims, represents "the culmination of money's evolution from an asset form to a debit form," and thus an unprecedented expression of capitalist megalomania (Hudson 2005, 17).

For Marx this evolution is not so much a chronological development as a tendency intrinsic to the capitalist mode of production from the beginning. In this sense, and although there have been all kinds of institutional limits to the historical reproduction of debt, capital represents that mode of production in which the debt form strives to liberate itself from all mediation, in space and in time. "[A]s representative of the general form of wealth—money—capital is the endless and limitless drive to go beyond its limiting barrier. Every boundary [*Grenze*] is and has to be a barrier [*Schranke*] for it" (Marx [1857] 1993, 334).[22] Nevertheless, if there is something that distinguishes the contemporary debt form, it is not simply its paradoxical relationship to U.S. imperial power, but also the level of production at which it operates. What is at stake in the accumulation of capital today is the regeneration of the biosphere—that is, the limits of the earth itself.

This is not merely an economic phenomenon then; it is also ecological. Paraphrasing the political economists, one might thus go further and argue that we live in a world in which the debt form is no longer referenced to any known terrestrial reserves, at least as far as the current state of science is concerned. The promise of capital in its present form—which after all is still irresistibly tied to oil—now so far outweighs the earth's geological reserves that we are already living on borrowed time, beyond the limits. U.S. debt imperialism is currently reproducing itself with an utter obliviousness to the imminent depletion of oil reserves. Fueling this apparently precarious situation is the delirium of the debt form, which in effect enables capital to reproduce itself in a realm of pure promise, in excess of the earth's actual limits, at least for a while. This is a delirium that operates between the poles of utter exhaustion and manic overproduction, premature obsolescence and the promise of surplus. In the sense that the debt can never be redeemed once and for all and must be perpetually renewed, it reduces the inhabitable present to a bare minimum, a point of bifurcation, strung out between a future that is about to be and a past that will have been. It thus confronts the present as the ultimate limit, to be deflected at all costs.

The speculative moment is only one side of the debt form, however, since the debt needs at some point to redeem its promised futures, to remember them to the past as if they had already been realized. In this way the debt form is not merely promissory or escapist but also deeply materialist; that is, it seeks to materialize its promise in the production of matter, forces, and things. In the long run what it wants to do is return to the earth, recapturing the reproduction of life itself within the promissory accumulation of the debt form, so that the renewal of debt coincides with the regeneration of life on earth—and beyond. It dreams of reproducing the self-valorization of debt in the form of biological autopoiesis.

BIOLOGY BEYOND THE LIMITS: DESTANDARDIZING LIFE

It is no accident that within biology itself, over the same period, the implicit understanding of what constitutes biological reproduction has been undergoing a rapid expansion. In the process the notion of what is technically possible in the reproduction of life has been similarly pushed to the limits. In this respect, and despite continuities, a rough distinction can be drawn between the dominant modes of biological reproduction that were developed in the first part of the twentieth century and the emerging technologies of the late twentieth and early twenty-first centuries. From plant selection and hybridization to animal reproductive medicine, modern biotechnology was primarily concerned with the industrial-scale reproduction of standardized life forms (Kloppenburg 1988; Clarke 1998). Each of these sciences drew in one way or another on the Weismannian paradigm of germinal transmission, in which the heritable essence of an organism is transmitted, through sexual recombination, from one generation to the next. Together with the insights of Mendelian genetics, and the scaling up of these processes effected by population genetics, the Weismannian approach seemed to suggest that biological reproduction, like any other area of the Fordist economy, could be subject to the demands of mass, standardized manufacture. Yet at the same time biological reproduction per se remained marginal to the prime sites of industrial production, which were much more attuned to the possibilities of the geochemical sciences (petrol- and chemical-based agriculture) than the life sciences.

When viewed in this context, a striking feature of the more recent life sciences (taking the invention of recombinant DNA in 1973 as a rough turning point) is their tendency to challenge the limits established by both the Weismannian-Mendelian paradigm and the industrial mode of biological reproduction. At issue here, often quite explicitly, is the question of the limits in which life can reproduce, regenerate, or simply survive. Is the vertical transmission of genetic material the principle mode of biological generation, as the Weismannian paradigm and Mendelian genetics would seem to suggest? Is microbial life phylogenetically inferior to the germinal life of organisms, with its commitment to species boundaries and sexual reproduction? Moreover, what are the ecological boundaries to the proliferation of life? Is life limited to the outer surface of the earth, normative atmospheric and biochemical conditions, and certain strict parameters of temperature and pressure? Is life limited to the earth itself?

Not surprisingly, many of these questions are being posed at the margins or even outside of the life sciences by such disciplines as physics, geology, and even cosmology. Is life constrained by the geochemistry of the earth, or is it capable of transforming these conditions in ways that fundamentally alter the course of evolution? Do the laws of physics dictate the parameters in which life is capable of unfolding? What become of the laws of physics (in particular, the second law of thermodynamics) when confronted with the growing awareness of life's powers of metabolic and environmental transformation? What are the implications for industrial production? And more speculatively, what are the implications for our understanding of the evolution and future of the earth? Is the history of the earth more accurately recounted in biospheric rather than geochemical terms? As one scientific review astutely put it, the notion of life itself is undergoing a dramatic destandardization such that the life sciences are increasingly looking to the extremes rather than the norms of biological existence (Rothschild and Mancinelli 2001, 1092). Importantly, these new ways of theorizing life are never far removed from a concern with new ways of mobilizing life as a technological resource.

This process of destandardization is perhaps most visible in the invention of recombinant DNA, the technique that is credited with having initiated the genetic revolution. In essence, recombinant DNA (or genetic engineering) is a method that allows biologists to generalize the processes of bacterial recombination to the whole of organic life. Bacteria are able to exchange mobile elements of genetic information among themselves, and it is these elements (or vectors) that genetic engineering makes use of in order to create chimeric organisms. Thus where traditional breeding methods are limited by the rule of sexual compatibility, recombinant DNA allows biologists to move sequences of genetic information across the barriers of species and genus, transferring DNA from plants and animals to bacteria and back again.[23]

Recombinant DNA (rDNA) differs from previous modes of biological production in a number of ways. First, while microbial biotechnologies such as fermentation are among the oldest recorded instances of biological production, recombinant DNA constitutes the first attempt to mobilize the specific reproductive processes of bacteria as a way of generating new life forms.[24] Moreover, recombinant DNA differs from the industrial mode of plant and animal production in the sense that it mobilizes the transversal processes of bacterial recombination rather than the vertical transmission of genetic information. This is a technique that lends itself to the specific demands of post-Fordist production—flexibility and speed of change—to a degree that was impossible in traditional plant breeding.

At the same time the production of transgenic organisms as viable life forms is having an interesting effect on theoretical understandings of the relationship between microbial and germinal or organismic life. The large-scale cultivation of transgenic crops has spurred a new research interest in bacterial recombination, and the results of this research have in turn demonstrated that horizontal gene transfer is much more extensive in nature than previously thought (Miller and Day 2004). The theoretical biologists Lynn Margulis, James Lovelock, and Dorion Sagan accord so much importance to the transversal reproductive processes of bacteria that they propose rereading the evolution of life as such from the point of view of the microcosm (Lovelock 1987; Margulis and Sagan 1997). For these theorists, it would seem, microbes represent the most *lifelike* of living organisms, precisely because they are so indifferent to the limits that constrain the reproduction of whole organisms.

In other areas as well the biosciences are pushing up against the limits within which life was previously confined. In addition to research into novel modes of microbial generation, scientists are also discovering that organisms, especially such microbes as archaea or bacteria, are able to survive extreme environments once considered off-limits for organic life (Ashcroft 2001; Rothschild and Mancinelli 2001). The so-called extremophiles are microbes that tolerate and even flourish under extreme geochemical and physical conditions. Microorganisms have been found deep beneath the earth's crust and in the depths of the ocean, in spaces once considered inhospitable to all forms of life. Others have been found to tolerate extremes of pressure, temperature, salinity, pH, and even radiation. Moreover, where it was once assumed that life depended on both oxygen and light, it has become apparent that certain microorganisms are able to live in the absence of both, and instead use manganese, iron, and sulphur to break down rocks into sources of food.

The discovery of extremophiles has opened up new lines of research in organic chemistry (it is not known, for example, exactly how the protein structures of certain microorganisms are able to survive temperatures close to boiling). It has also thrown into question received wisdom about the relationship between the biological and geochemical—after all, if microorganisms are able to metabolize and transform inorganic matter into organic compounds, it becomes legitimate to inquire into their role in the geological evolution of the earth. Does life adapt to environmental niches, or actively incorporate and transform them? It is typical of the anticipatory dynamics of the contemporary biotech industry that even while such theoretical questions remain open, efforts are already under way to translate extremophile research into commercial outcomes. The already voluminous literature in the field predicts that extremophile research will form the basis for a second wave in bioremediation technologies (a method that uses microorganisms to degrade or transform toxic waste products) (Watanabe 2001).

The intellectual context for much of this work can be found in the concept of the "biosphere," first developed by the Russian scientist Vladimir Vernadsky (1863–1945) and later taken up by the British atmospheric chemist and inventor of the Gaia hypothesis, James E. Lovelock, while he was working for the NASA space program in the 1960s. For Vernadsky life itself is the most significant determining influence in the evolution of the earth and its atmosphere. The study of geochemistry and its laws therefore needs to be integrated into a larger synthesis—that of *bio*geochemistry. At the same time, what Vernadsky understands by life is something much more comprehensive than the commonly received definitions of the biosciences themselves. Life is biospheric: it encompasses the ensemble of biota inhabiting the planet, from the microbial to the human. It is self-regulating or self-responsive; far from adapting to external, geochemical conditions, the evolution of any one life form is in the first instance determined by its relations to other life forms. And considered at a biospheric level, the defining feature of life is its ability to transform solar radiation into new chemical compounds, thus accumulating a relentless surplus of "free energy" over and above the chemical deposits already available on earth (Vernadsky [1929] 1998, 50–60).

In this way what characterizes the time arrow of biospheric evolution is not the progressive exhaustion of differences, but rather the continuous disturbance of geological, chemical, and atmospheric equilibria, via which life renews the chemical imbalances of the earth. Life, in this view, is intrinsically expansive—its field of stability is neither rigorously determined nor constant (ibid., 113). Its law of evolution is one of increasing complexity rather than entropic decline, and its specific creativity is autopoietic rather than adaptive.

When Lovelock, in collaboration with the microbiologist Margulis, takes up the thesis of the biosphere, they relocate the driving force of evolution in the world of microbes rather than plant life. Availing themselves of recent research into microbial recombination and Margulis's own theory of symbiosis, Margulis and Lovelock argue that the "microcosm" is the prime mover of biospheric evolution. The central thesis of their work (the so-called Gaia hypothesis) is that the biosphere's capacity to regulate itself is crucially dependent on the metabolic processes of microorganisms. From this starting point, Margulis and Lovelock develop a philosophy of waste and regeneration profoundly at odds with the thesis of ecological limits to growth. Responding to calls for greater regulation of pollution, Lovelock (1987, 27) asserts that the production of waste is an inevitable consequence of life's cycles of energy transformation: "Pollution is not, as we are often told, a product of moral turpitude. It is an inevitable consequence of life at work. The second law of thermodynamics clearly states that the low entropy and intricate, dynamic organization of a living system can only function through the excretion of low-grade products and low-grade energy to the environment."

The production of waste is so inescapable, Lovelock's collaborator Margulis later argues, that the history of microbial evolution should be read as a succession of catastrophic pollution events, many of them much greater than the contemporary threat posed by industrial waste (Margulis and Sagan 1997, 99–114). Yet for these theorists the accumulation of waste products, although fatal to particular life forms, will never be enough to stop the evolution of *life itself*. Indeed, the continuing evolution of life—life's very capacity to innovate—is intimately dependent on periodic waves of environmental crisis. Life creates its own limits to growth only to expand beyond them: "It is an illuminating peculiarity of the microcosm that explosive geological events in the past have *never* led to the *total* destruction of the biosphere. Indeed, like an artist whose misery catalyzes beautiful works of art, extensive catastrophe seems to have immediately preceded major evolutionary innovations. Life on earth answers threats, injuries, and losses with innovations, growth and reproduction. . . . With each crisis the biosphere seems to take one step backward and two steps forward—the two steps forward being the evolutionary solution that surmounts the boundaries of the original problem" (ibid., 236–37).

In the work of Margulis, Lovelock, and Sagan biosphere science remains largely propositional and intuitive. Other theorists, particularly those associated with the sciences of complexity, have developed a more solid mathematical framework for thinking through the dynamics of self-organizing systems. The work of Ilya Prigogine and Isabelle Stengers (1979, 1984, 1992) in particular has gone furthest in exploring the implications of biosphere science for understanding the relationship between the physical-chemical laws of nature and the specific temporality of life. The later, more philosophically oriented work that Prigogine and Stengers undertook together has its origins in Prigogine's earlier research into dissipative structures, where he was interested in questioning the universality of the law of entropy for describing the evolution of physical systems. Far from constituting a universal law of nature, Prigogine argued, the second law was valid only for closed systems, those that were cut off from exchanges of matter and energy with the outside world.

Turning his attention instead to the evolution of open or dissipative structures, Prigogine sought to show that the apparently wasteful consumption of matter and energy (dissipation or waste production) can lead to the creation of highly organized, complex, evolving structures, rather than the inevitable depletion of possibilities prescribed by the second law. Indeed, his work suggests that the complexification of structure is the rule rather than the exception for open systems. The law of entropy remains valid for closed worlds, but increasing complexity is the overriding tendency in dissipative structures. And providing we enlarge the conditions of experimentation sufficiently, complexification is without predetermined limit: a dissipative structure evolves through successive thresholds of disequilibrium, at which point it is impelled to bifurcate along one of several paths of organization, none of which can be predicted from initial conditions.

The context for Prigogine's independent work is the physical sciences, thermodynamics and chemistry, but even here he makes special reference to biological processes as the most appropriate models for complexity in nature. Hence, "we see instability, fluctuation and evolution to organized states as a general non-equilibrium process whose most spectacular manifestation is the evolution of life" (Prigogine and Kondepudi 1998, 452). And in terms very close to those of biosphere science, Prigogine conceives of the earth as an open, dissipative system regulated and maintained far-from-equilibrium by the continuing evolution of life (ibid., 409).

In their coauthored work, Prigogine and Stengers insist on the significance of complexity theory for re-visioning the relationship between the physical and biological sciences. While nineteenth-century science tended to see biological evolution as an exception to the general laws of nature—on the pretext that the creation of diversity runs counter to the second law—complexity theory reverses the order of priority between the physical and the life sciences, so that the complexification of life claims the status of universal law. Far from representing the exception to the usual laws of physics and geochemical nature, the ontogeny and evolution of life here become the paradigm of dynamic processes in general: "In the context of the physics of irreversible processes, the results of biology . . . have a different meaning and different implications. We know today that both the biosphere as a whole as well as its components, living or dead, exist in far-from-equilibrium conditions. In this context life, far from being outside the natural order, appears as the supreme expression of the self-organizing events that occur" (Prigogine and Stengers 1984, 185).

Importantly, Prigogine and Stengers (ibid., 203–9) insist that the science of complexity calls for a "new political economy of nature," one that would look to the laws of biological development and evolution rather than the industrial machines that fueled the economy of the nineteenth century. Their assumption here is that the productivity of biological processes is fundamentally different from that of inorganic nature: while industrial machines are subject to the laws of depletion and diminishing returns, life at its most "lifelike" obeys a law of self-organization and increasing complexity. Where industrial production depends on the finite reserves available on planet earth, *life, like contemporary debt production, needs to be understood as a process of continuous autopoiesis, a self-engendering of life from life, without conceivable beginning or end.* Thus in the work of Prigogine and Stengers the time arrow of life comes to represent a general principle of complexification, running counter to the Malthusian thesis of ultimate limits to growth. "There can be no end to history," Prigogine has claimed (Prigogine and Nicolis 1989, 126). What this absence of limits signifies, however, is not so much a progressive, exponential arithmetic of growth as a fractal one (Prigogine and Stengers 1992, 72–74).[25] The complexification of life is without end in the sense that it evolves toward no finite limit or equilibrium point.

In this way the growing influence of complexity theory across the life sciences has ushered in a new emphasis on catastrophism in the theory of evolution. In place of Darwin's gradualism—itself dependent on a slow vision of geological time—evolutionary theorists are increasingly interested in the catastrophe events that punctuate evolution in the form of mass extinctions. For these theorists life no longer merely inhabits the geological extremes; it also survives a succession of extreme catastrophe events. Its moments of crisis are inevitable but strictly incalculable: like fractal discontinuities they are not subject to the standard distribution of statistical events (Bak 1996; Kauffman 1995 and 2000). Indeed, in the work of someone like theoretical biologist Stuart Kauffman, the catastrophe event becomes the very condition of life's continuing tendency toward complexity: life evolves through periodic moments of crisis; the creation of new life, of biological innovation, requires the perpetual destruction of the old.

At this point it becomes legitimate to ask what is meant by the term "life" in these theories. For biosphere theorists Margulis, Lovelock, and Sagan, life manifests itself most powerfully in the relentlessly divisible, mutable world of microbes. Their rereading of evolution thus concludes with the certitude that microbial life will outsurvive all limits to growth—certainly it will outsurvive the human race and quite possibly the end of the earth: "Neither the existence of species nor species extinction is a property of bacteria. Although individual bacterial death is continuous, fierce pressures on the moneran kingdom as a world-wide gene-exchanging enterprise led to the rapid exchange of natural biotechnologies, enormous population growth rates, and in general the ability to persevere with metabolic talents intact even during the most severe planetary crisis" (Margulis and Sagan 1997, 275).

Ultimately, however, the work of Prigogine and Stengers goes even further and transforms life into a biocosmological principle—a universal time arrow. What the complexity turn demands, they claim, is a whole new approach to cosmology, one that looks to the self-organizing processes of biological evolution as a key to universal time. Prigogine and Stengers thus replace Lord Kelvin's nineteenth-century cosmology of decline—the second law as a force of entropy leading to the heat death of the universe—with a biocosmological law of increasing complexity. Here there is no beginning or end to cosmological time, but rather a series of catastrophic bifurcation events out of which universes continuously rebirth each other. In the work of Prigogine and Stengers, who are not astrophysicists, these cosmic visions are necessarily presented in the manner of philosophical deduction—but complexity is also beginning to exert a real influence in cosmological theory itself. As the controversial cosmologist Eric Lerner (1991, 394) has recounted in his study of recent directions in the field: "Cosmologically, a universe with as little matter as ours will never collapse. Nor does thermodynamics even demand that the universe run down: Prigogine has demonstrated that there is no inherent limit to the order the universe will attain, or to its increasing energy flows. Our universe is speeding away from the 'heat death' of total equilibrium."

Cosmological musings of this kind may appear to belong to the outer orbits of speculative science. But they also inform much of the thinking behind NASA's very results-oriented space biology program. NASA has in fact funded research into exobiology since the 1960s, the period when Lovelock was first hired, and famously sent two Viking explorer missions to search out life on Mars in 1976. The area, rebaptized "astrobiology," took on a new lease of life in 1995 when NASA was forced to restructure as a result of funding cuts.[26] The restructuring could conceivably have led to the loss of the whole space biology program, but instead, with the help of an administrator well-disposed to the possibilities of the biotech revolution, the area was reprioritized, endowed with a newfound commercial ethos, and summonsed to formulate an ambitious vision of its future research directions.

This vision was outlined by 150 scientists in the course of a series of conferences held in the late 1990s, culminating with the publication of a twenty-year Astrobiology Roadmap in January 1999. The roadmap reiterated two of the original goals of the NASA exobiology program: to search for life on other planets and to understand the origins of life on earth (Dick and Strick 2004, 227–29). But it also added a new angle, which significantly transformed the practical and political consequences of NASA's work. Henceforth, the astrobiology program would not only seek to detect the presence of actual or past life on other planets, hoping to use such findings to reflect back on their significance for the history of life on earth. It would now be required to concern itself with the *future* evolution of life, both on earth and elsewhere (ibid., 231). To this end, NASA's Ames Research Center has become increasingly involved in designing and funding research into extreme life forms and microbial bioremediation. Work on earthly extremophiles, it is surmised, might provide clues to the possibilities of life on other apparently inhospitable planets. But it might also reflect back on the possibilities of life's continued survival on earth. Implicit here is the whole problematic of ecological crisis and the threat it poses not only to human existence but to the biosphere itself.

The NASA research program addresses these problems not only in a predictive mode (is survival more or less probable?), but also with a view to preemptive technological intervention. The question is not only, Is life reproducible at certain atmospheric and geochemical extremes? But in the last instance, How is it possible to make life survive the extremes? In other words how can we re-create life beyond the limits? In an important sense, then, astrobiology has moved beyond the speculative questions of its predecessor, exobiology, to become technically interventionist. Moreover, it takes the biotech revolution one step further in that it is not only interested in the reproduction of particular life forms but, more ambitiously, in *terraformation*—the creation of entire life worlds or biospheres. In their historical account of the NASA astrobiology program, Steven Dick and James Strick elaborate on the importance of this shift:

> The future of life on earth and beyond—a question hardly enunciated in early exobiology—remained the most underdeveloped of astrobiology's . . . questions. Many scientists were not accustomed to dealing with the future. . . . Nevertheless, precisely because of the lack of attention, the potential for new thinking and important discoveries was great. As the astrobiology roadmap had stated, NASA had much to contribute to global problems such as ecosystem response to rapid environmental change and Earth's future habitability in terms of interactions between the biosphere and the chemistry and radiation balance of the atmosphere. It was uniquely suited to understanding the human directed processes by which life could evolve beyond earth. . . . Problems such as terraforming Mars were indeed problems of the future but no less important for that. . . . NASA's vision for the future was "to improve life here. To extend life there. To find life beyond." (ibid., 230)

The importance of the NASA space biology program lies in the translational role it is playing between speculative science and (post)industrial applications. Over recent years NASA has become increasingly present in funding and initiating research and development in the areas of extremophile science, bioremediation, and alternative fuel technologies. And as Margulis points out, even the Gaia hypothesis is finding its way into academic science programs and funding proposals, now that it has taken on the more respectable guise of astrobiology and earth systems science (Margulis 2004).

In light of this trend toward institutionalization, the political and ecological consequences of biosphere and complexity science are becoming difficult to ignore. Such theories may well have their origins in essentially revolutionary histories of the earth (Davis 1998, 15–16), but in the current context they are more likely to lend themselves to a distinctly neoliberal antienvironmentalism (Buell 2003). Whether this is a misinterpretation of complexity theory, at odds with the intentions of the theorists themselves, is in a sense beside the point, since in the absence of any substantive critique of political economy, any philosophy of *life as such* runs the risk of celebrating *life as it is.* And the danger is only exacerbated in a context such as ours, where capitalist relations have so intensively invested in the realm of biological reproduction. Even in the work of Prigogine and Stengers the new political economy of nature sounds suspiciously like the new political economy of neoliberalism. And although their critique of limits-to-growth theories is logically impeccable, they offer little else in the way of practical political alternatives than the reassurance that life itself—life in its biospheric and even cosmic dimensions—will ultimately overcome all limits to growth.

In the work of James Lovelock such certitudes are combined with a blatant stance against environmental regulation of any kind, culminating most recently in his public defense of nuclear energy as the answer to the imminent depletion of fossil fuels. But perhaps it is not so much the implicit antienvironmentalism of these theorists that is remarkable in itself (it is after all of a piece with the ubiquitous free-market critique of state regulation) as the fact that it stems from a position that can only be described as vitalist. It is because life is neguentropic, it seems, that economic growth is without end. And it is because life is self-organizing that we should reject all state regulation of markets. This is a vitalism that comes dangerously close to equating the evolution of life with that of capital.

The theoretical biologist Stuart Kauffman (1995, 208–9) makes the link quite explicitly, bringing together classical liberal theories of self-regulating economic growth with a newfound catastrophism: "Adam Smith first told us the idea of an invisible hand in his treatise *The Wealth of Nations.* Each economic actor, acting for its own selfish end, would blindly bring about the benefit of all. If selection acts only at the level of the individual, naturally sifting for fitter variants that 'selfishly' leave more offspring, then the emergent order of communities, ecosystems, and coevolving systems, and the evolution of coevolution itself are the work of an invisible choreographer. We seek the laws that constitute that choreographer. And we will find hints of such laws, for the evolution of coevolution may bring coevolving species to hover ever poised between order and chaos, in the region I have called the edge of chaos."

"On a larger scale," Kauffman (ibid., 296–97) continues, "persistent innovation in an economy may depend fundamentally on its supracritical character. New goods and services create niches that call forth the innovations of further new goods and services. Each may unleash growth both because of increasing returns in the early phase of improvement on learning curves or new open markets. Some of these are truly triggers of Schumpeterian 'gales of creative destruction,' ushering out many old technologies, ushering in many new ones in vast avalanches. Such avalanches create enormous arenas of increasing returns. . . . Diversity begets diversity, driving the growth of complexity." This is a philosophy that celebrates capitalism as a catastrophic life principle, a biological and economic law of crisis-ridden yet relentless growth. And given the neoliberal sensibilities of such theoretical biologists as Kauffman, it should come as no surprise that there is a growing interest in complexity theory among economists themselves. Just as complexity theorists are celebrating the new political economy of nature, a certain kind of vitalism is coming back into fashion in economics.

GROWTH BEYOND THE LIMIT: THE NEW LAISSEZ-FAIRE

After decades in which economics was dominated by the equilibrium models of classical mechanics, the biological influence is returning with a vengeance in new theories of economic growth. This influence is evident in such diverse contexts as the new academic approaches of evolutionary economics, the populist new economy literature espoused by the likes of *Business Week* and *Wired*, and the work of journalists Michael J. Mandel, Kevin Kelly, and George Gilder.[27] In one sense this trend could be interpreted as a return to classical liberal models of growth, which were arguably more influenced by biological and evolutionary theories than by mechanics. This is a return with a difference, however. The classical liberal economist Adam Smith famously envisaged the economy as a multitude of laboring forces evolving from one equilibrium state to another—and it is this uniformitarian, steady-state vision of growth that also informs Darwin's political economy of nature.

The new liberal economists, however, are more likely to theorize growth as a process of evolution in far-from-equilibrium conditions. These economists remain true liberals, in the sense that they believe in the essential autonomy of the market: its capacity to self-organize. Yet in place of Adam Smith's principle of equilibrium—the invisible hand of the self-regulating economy—they argue that economies evolve most productively in far-from-equilibrium conditions. What is *neo* about neoliberalism is its tendency to couple the idea of the self-organizing economy with the necessity for continual crisis (a conceptual move that has obvious affinities with complexity theory). A first step in this direction was made by arch neoliberal theorist Friedrich von Hayek as early as 1969, when he decided to trade in his self-organizing equilibrium models of the economy for biological models of nonlinear development. He wrote: "Even such relatively simple constituents of biological phenomena as feedback (or cybernetic) systems, in which a certain combination of physical structures produces an overall structure possessing distinct characteristic properties, require for their description something much more elaborate than anything describing the general laws of mechanics" (Hayek 1969, 26).

It was only with the dissemination of mathematical models associated with complexity theory, however, that such ideas would attain a certain degree of credibility among economists (Mirowski 1997). And one institution in particular played a key role in facilitating this process—the privately funded Santa Fe Institute of California. Throughout the 1980s the institute became an intense site of conceptual exchange between economists, theoretical biologists, and evolutionary theorists, all of whom were interested in developing the insights of nonlinear, complex systems theory for thinking through processes of growth of one kind or another. In 1987 a first conference on "The Economy as a Complex Evolving System" was held, followed by a second one in the late 1990s (Anderson, Arrow, and Pines 1988; Arthur, Durlauf, and Lane 1997).[28] The conferences brought together economists and natural scientists of various denominations (notable participants included Stuart Kauffman and innovation economists Brian Arthur and Kenneth Arrow). Despite these differences, what emerged from the proceedings was an overwhelming amount of consensus.

For all these theorists, it seems, the complexity approach to economic and/or biological evolution entails several basic presuppositions: first, complex systems evolve best in far-from-equilibrium conditions or at the edge of chaos, to adopt Kauffman's phraseology; moreover, such systems evolve most productively when they are free from external regulation—complex systems in other words prefer to self-organize; and finally, although an individual complex system eventually exhausts its possibilities of further differentiation, there is no essential limit to the evolution of complexity per se. In nature as in economics the law of complexity is one of increasing returns punctuated by periodic moments of crisis.

Perhaps the one common point of reference for complexity theorists in the natural sciences and economics is the work of the Austrian economist Joseph Schumpeter, who developed an evolutionary theory of innovation economies at a time when biological models were definitely not in vogue. For Schumpeter (1934) only biological models of growth could provide economists with the necessary tools for thinking through the historicity of economic dynamics. But his perspective on evolution was also curiously out of step with the dominant, uniformitarian theories of his time. The evolution of both life and capital, as Schumpeter envisaged it, was punctuated by violent but ultimately productive moments of crisis, which he described as "creative destruction." In fact, the thesis of creative destruction in many ways anticipates the new catastrophism of evolutionary theory itself and already transposes it into the realm of economic life. For Schumpeter the time arrow of innovation is necessarily "convulsive," the unfettered growth of capital subject to violent booms and busts. Again, it is not surprising to see complexity theorists in the natural sciences, from Prigogine and Stengers (1984, 207–8) to Kauffman (1995, 296–7), referencing the work of Schumpeter as if it provided a model for the evolution of life itself.

INDUSTRIALISM BEYOND THE LIMIT: BIOREMEDIATION, ENERGY FUTURES, AND THE BIOECONOMY

Although theories of complex biological and economic growth have largely developed in parallel academic universes, there is one discourse that merges them together—that of the bioeconomy. In 2005 the OECD launched a two-year project aiming to "draft a broad policy agenda for governments in respect to the bioeconomy" (OECD 2005, 1). Noting that the bioeconomy is a "new concept," the OECD proposed a definition that nicely brings together the possible alliances between biological productivity and the extraction of surplus-value: the bioeconomy is defined as that part of economic activities "which captures the latent value in biological processes and renewable bioresources to produce improved health and sustainable growth and development" (ibid., 5). The concept of the bioeconomy, however, has a longer history than the OECD report would seem to suggest, and this history is very revealing about the specific political interests that inform it.

One of the areas in which the premise of the bioeconomy was first developed was the environmental sciences. In *The Politics of Environmental Discourse:*

Ecological Modernization and the Policy Process, Maarten A. Hajer outlines the emergence of a new way of thinking about ecological crisis, one that purports to offer an alternative to the regulationist, antigrowth recommendations of the Club of Rome. The contention of ecological modernization is that "economic growth and the resolution of ecological problems can . . . be reconciled" without the need for prohibitive regulation (Hajer 1995, 26). Instead, it proposes a range of positive incentives designed to encourage industries to *internalize* ecological limits within their accounting strategies, so that environmental solutions become economically attractive. What this requires is a translation of "the discursive elements derived from the natural sciences" into monetary signs, so that the two become mutually convertible (ibid.). In particular, ecological modernization encourages the use of future-oriented investment of the kind we have seen, for example, in pollution emissions trading or the sale of ecological futures on the Chicago stock market.[29] The interest of such innovations, according to one account, is to "create value where none existed before," in much the same way as the issuing of debt comes to function as a creation ex nihilo, "valued less on current earnings than on future potential" (Daily and Ellison 2002, 22).

Other environmental scientists go still further and recommend the adoption of strategies for incorporating biological growth into the very infrastructure of production. In their enormously popular work *Natural Capitalism*, Paul Hawken, Amory Lovins, and L. Hunter Lovins (1999, 73) outline their vision of an economic future in which the specific ability of life to self-regenerate—to transform "detritus into new life"—would be mobilized as a means of overcoming the waste-products of industrial production. Of course, the logic at work here was already evident in the early days of recombinant biotechnology, where all kinds of attempts were made to biologize petrochemical and pharmaceutical production: from the use of modified microorganisms to manufacture chemical substances to mammalian pharmaceutical factories, engineered to express medically useful human proteins in their milk or blood, plant production of plastics, and bioremediation (the use of modified microorganisms for cleaning up oil spills and toxic waste conversion).

Natural Capitalism, however, envisages a future in which these specific examples of bioproduction would form part of a generalized bioeconomy, one in which "biomimicry . . . inform[s] not just the design of specific manufacturing processes but also the structure and function of the whole economy" (ibid.,

73). Here the insights of biosphere theory come together with the growth imperatives of the new economics to suggest that the bioeconomy will take us beyond all limits, transforming even industrial waste into a source of surplus value. As the authors of *Natural Capitalism* succinctly put it, the "word [resource] comes from the Latin *resurgere*, to rise *again*" (ibid., 196).

It would be a mistake to think that such speculative solutions to ecological crisis were merely a product of the Clinton/Gore era, since in fact many of their most futuristic recommendations have been translated into explicit policy objectives under the notoriously antienvironmental Bush regime. In its 2004 twenty-year strategic research plan, for example, the Department of Energy's Office of Science foresees the progressive development of bioprocesses to clean and protect the environment and to provide new energy sources (U.S. Department of Energy 2004, 33; Carr 2005). And adopting a language that recalls the Gaia hypothesis as much as the more economistic calculations of ecological modernization, the report looks to the history of microbial and biospheric evolution as a source of future solutions to the looming energy crisis: "Over billions of years of evolution, Nature has created life's machinery—from molecules, microbes and complex organisms to the biosphere—all displaying remarkable capacities for efficiently capturing energy and controlling precise chemical reactions. The natural, adaptive processes of these systems offer important clues to designing solutions to some of our greatest challenges. . . . Such capabilities will provide us unprecedented opportunities to forge new pathways to energy production, environmental management, and medical diagnosis and treatment" (U.S. Department of Energy 2004, 33).

The report places special emphasis on the potential industrial applications of extremophiles. Research in the field has so far produced enzyme-based bioprocesses ranging from the banal (detergent additives) to the fundamental (PCR, or polymerase chain reaction, an important tool in basic biomedical research and diagnostics), and there is much speculation about using extremophiles for toxic-waste remediation. In recent years the field has moved far enough from the margins to the center of scientific inquiry that institutions ranging from the U.S. National Science Foundation to the Department of Energy and NASA have all injected sizable funds into the area. No less a figure than Craig Venter (the founder of Celera Genomics) is currently engaged in DOE-supported work on modified extremophiles as an alternative energy source.

But it is the U.S. Energy Policy Act of 2005 that has most decisively placed the question of the bioeconomy on the political agenda (Carr 2005). In seeming contradiction with the overall thrust of Bush's foreign policy, the act calls for a reduction in American dependence on foreign oil supplies and the promotion of domestic R & D in the production of alternative, bio-based fuels—a move that has been interpreted as a significant boost to the field of white or industrial biotechnology. With petroleum companies reporting a declining rate of new oil-reserve discoveries and industry analysts predicting that peak oil production has already been reached, the 2005 energy act represents a belated attempt by the U.S. government to come to terms with that ultimate limit to industrial growth—oil depletion.

As such, it is certainly more of a response to the increasingly visible strategic and economic costs of oil dependence than a sudden realization of its ecological consequences (to which capitalist modes of calculus are inherently blind in any case, according to Marx). And undoubtedly, the plan to divert at least part of federal investment from fossil fuels to bio-based fuels offers a prospective escape route from many of the United States' most pressing economic problems. These include, first, the rising costs of defending its dependence on Middle Eastern oil supplies; the growing competitive pressures from India and China, as they too become major consumers of world oil; and finally, both the relative failure of GM food production and the diplomatic costs of sustaining export subsidies to the U.S. farming industry. Moreover, the long-term goal of reorienting American agriculture toward the production of fuel crops represents one possible way of trumping China's rising economic power.

Here, U.S. industrial and foreign policy comes together with the speculative solutions developed in the context of the NASA space program to suggest ways in which America might quite literally remake the imperialist world—beyond the limits of the geochemical paradigm and its increasingly visible signs of depletion. As for Reagan's "Star Wars" program and the beginnings of biotech, the enormous leverage commanded by the United States' position at the center of world financial flows is what makes this delirium even thinkable. Fueling the dream that life will self-regenerate de novo is the momentum of the U.S. debt cycle, which must also maintain itself in a state of continuous and precarious renewal.

As an environmental strategy, however, the 2005 energy act is traversed by fundamental contradictions. At the same time as it purports to champion the cause of the emerging postindustrial bioeconomy, it does very little to reduce greenhouse gas emissions and dependence on fossil fuels. And while it rewards the powerful business interests behind ethanol and nuclear power production, it has eliminated support for research into geothermal, solar, and hydropower energy alternatives. Bush's environmental epiphany, moreover, sits uneasily with his refusal to sign up to the Kyoto protocol. This is a strategy that looks to the future to recuperate the costs of ecological depletion, while accelerating the actual production of wastes in the present. And insofar as the bio-based economy promises to regenerate waste—indeed to provide a solution for all of the limits associated with industrialism—it is in fact utterly dependent on the continuous expansion of waste production. Bush's energy act, in other words, appears as much designed to perpetuate ecological crisis as to overcome it.[30]

These contradictions, I suggest, are not simply the result of Bush's political incompetence. As long as life science production is subject to the imperatives of capitalist accumulation, the promise of a surplus of life will be predicated on a corresponding move to devaluate life. The two sides of the capitalist delirium—the drive to push beyond limits and the need to reimpose them, in the form of scarcity—must be understood as mutually constitutive. In one respect this is simply a restatement of Marx's reflections on the counterproductive tensions of capitalism. The difference today is "merely" that the tensions of capitalism are being played out on a global, biospheric scale and thus implicate the future of life on earth. It is therefore no coincidence that the dream of terraformation has arisen at a moment in history when capitalist modes of production are literally testing the limits of the earth. Nor is it incidental that the life sciences are promising to invent new forms of life at a time of accelerating extinction rates. The political problematic is twofold. How can we contest the depletion, extinction, and devaluation of living possibilities without opting for the wholesale capitalization of a surplus life to come? And how can we counter the relentless push to drive beyond the actual limits of the earth without sanctioning the politics of scarcity?

In a real sense we are living on the cusp between petrochemical and biospheric modes of accumulation, the foregone conclusion of oil depletion and the promise of bioregeneration. An effective ecological counterpolitics therefore needs to operate on both levels. In the United States there is already a considerable amount of organization around the politics of oil dependence. At the same time it is important to work in the prospective mode, to detect and preempt the new forms of scarcity that are being built into the promise of a bioregenerative economy. Today international relations theorists are openly contemplating a future in which environmental scarcity constitutes a major source of conflict and refugee movement. And the imminence of oil depletion is a millenarian scenario that animates both Islamic and evangelical fundamentalisms. It therefore becomes urgent to formulate a politics of ecological contestation that is neither survivalist nor techno-utopian in its solutions.

In recent years the preoccupation with biospheric limits to growth has been somewhat displaced: where the Club of Rome warned of impending limits to life on earth, and evolutionary theorists speculated about the end of the human race, the same phenomena are now more likely to be formulated in terms of human, biological, and environmental security.[31] The administration of biological scarcity, in other words, has moved from the realm of economic calculations to that of military concerns. In this chapter I have considered the dynamics of contemporary capitalism as they relate to the ecological problematic of the biosphere. In chapter 2, I turn my attention to the politics of infectious disease, biomedicine, and drugs, since it is in the arena of global public health that the neoliberal promise of a surplus of life is most visibly predicated on a corresponding devaluation of life. It is also here that the tendency to invoke the discourse of security is becoming increasingly apparent.

□ 2 □

ON PHARMACEUTICAL EMPIRE

AIDS, Security, and Exorcism

Exactly when they came to be known as *complex humanitarian emergencies* and who gave them that name is not clear, but by 1990 many senior disaster managers had begun using the term.
—Andrew S. Natsios, *U.S. Foreign Policy and the Four Horsemen of the Apocalypse*

THROUGHOUT THE 1980S A NEW CONSENSUS GRADUALLY TOOK HOLD OF the international political community. The AIDS epidemic was no longer a global public health issue (as UNAIDS had tried to establish under the leadership of Jonathan Mann) but rather the preeminent security threat of the twenty-first century. In all its sexual and social complexities, AIDS was to be confronted as a military emergency. The first step toward this reconfiguration of AIDS was made in 2000, when the UN Security Council, which had traditionally been oblivious to health issues, dedicated its inaugural meeting of the millennium to the growing impact of AIDS in Africa. At about the same time, and from an avowedly self-interested perspective, the United States commissioned two reports on the national security implications of infectious disease. In one report the National Intelligence Council warned that "new and reemerging infectious disease [would] pose a rising global health threat and [would] complicate U.S. and global security over the next twenty years" (National Intelligence Council 2000, 1), while a document jointly issued by the Chemical and Biological Arms Control Institute and the Center for Strategic and International Studies predicted that in the aftermath of the Cold War cross-border contagion was set to become the number-one international security risk (CBACI and CSIS 2000).[1] Both reports placed particular emphasis on the dramatic rise of HIV/AIDS infections in sub-Saharan Africa.

In retrospect, it could be countered that the transformation of AIDS into a security issue didn't come totally out of the blue. Rather, it was consistent with a wider redefinition of strategy that throughout the 1990s, tended to merge war with humanitarian intervention and public health crises with military emergencies. Crucial to these developments was the growing consensus that the paradigmatic military threat of the post–Cold War era was no longer to be found in the formal declaration of war between sovereign states but rather in the irruption of so-called complex emergencies—natural or human-made disasters characterized by the implosion of the state, the breakdown of essential public infrastructures (sanitation, water, power, and food supplies), and the prevalence of infectious disease.[2]

For infectious disease was undoubtedly back, despite the long-held belief that the modern pharmaceutical industry would see to the near total elimination of infection-related deaths by the beginning of the millennium. Nowhere was the stark discrepancy between the promise of biomedicine and the obstinate, inexplicable endurance of infectious disease more apparent than in the case of the AIDS pandemic. The development of AIDS drugs in the late 1980s, followed by the powerful antiretroviral therapies of the mid-1990s, had seemed to signify that AIDS too, the newest of infectious diseases, would succumb to the inexorable advance of the miracle drug. In the richest countries of the world, historians of the disease were proclaiming that AIDS had morphed into a chronic condition like any other, to be taken in charge on a life-long basis by the pharmaceutical industry. But despite the blunt fact that AIDS mortality could now be significantly reduced, the promised health transition from death sentence to chronic condition failed to materialize in many developing countries. This is particularly true of sub-Saharan Africa, where the numbers of people infected with the disease has continued to escalate. In South Africa alone it is estimated that five million adults between the ages of fifteen and forty are infected with HIV/AIDS, the largest absolute number of any country in the world. Very few of these people have access to the new antiretroviral treatments.

Yet in the face of such self-evidence and despite the recommendations of its own security advisers, the United States has made every effort to *prevent* the South African government from designating the AIDS epidemic as a state of emergency. Beginning in the mid-1990s—the watershed decade that saw the end of apartheid and the introduction of antiretrovirals (ARVs)—the United States has consistently tried to dissuade South Africa from utilizing the emergency clause that would allow it to override World Trade Organization (WTO) rules on the importing of low cost generic drugs. The WTO, which since the mid-1990s effectively oversees international trade law on pharmaceuticals, does in fact acknowledge the exceptional status of such complex emergencies as the AIDS epidemic. Under Article 31 of the agreement on trade-related intellectual property rights (TRIPs), the use of patented drugs may be authorized "in the case of a national emergency or other circumstances of extreme urgency."[3] Escape clauses of this kind are not unusual in patent legislation; the U.S. government itself enjoys far wider powers of compulsory licensure and is capable of overriding monopoly rights virtually at will.[4] Yet it was with the full support of the United States and several European governments that in 1998 forty-one pharmaceutical companies brought the South African government to trial, naming Nelson Mandela as defendant.

In bringing the case, the drug companies were taking issue with the South African Medicines Act of 1997, which granted the health minister discretion to access affordable medicines through parallel-importing or compulsory licensing.[5] The act, it was claimed, was a transgression of the WTO accords on patent law. Not only did it deprive the pharmaceutical market of an emerging potential market; it threatened, through example, to spread the virus of patent violation to all the promising new markets of the developing world. Most alarmingly, perhaps, the South African government's intervention on behalf of public health threatened to call into question the exorbitant prices imposed by the pharmaceutical industry in its most lucrative of markets—the United States. Acting on behalf of its most profitable industry, the U.S. government threw its full weight behind the court case, warning of trade sanctions if the act was not repealed.

What can the recent history of the AIDS virus tell us about the neoliberal politics of innovation? How does neoliberalism attempt to determine the price of life? And how do these price-fixing strategies play out on a global scale? Here I am interested in exploring the flip side of the debt relationship that I outlined in chapter 1. The very juncture that enabled the United States to reconfigure its imperial power on the basis of a perpetually renewed debt at the same time brought impossible debt burdens to many countries of the developing world. I am therefore interested in the ways in which the macropolitics of debt servitude have impacted on the everyday micropolitics of bodies, creating new geographies of labor, sex, and contagion, and reworking the very epidemiology of disease. At the same time I explore the new discourse on security that has arisen in parallel with the debt crisis of many southern African nations and which in many ways ends up criminalizing the very social disruptions it has provoked. How did it become possible, even logical, to redefine AIDS as a security issue, indeed the preeminent security threat of the new millennium? And what are the implications for our understanding of both security and contagion? What does the discourse of human or biological security tell us about the nature of imperialist, biopolitical relations today—as compared, for example, with the biopolitics of the welfare state and developmentalism?

Thus far my critique has been directed against the imperialist and global interests that inform biomedical and pharmaceutical research today. But the AIDS epidemic in South Africa is also mired in the internal politics of the postcolonial state and its ambivalent relationship to its apartheid-era, public health history. Perhaps the most intriguing aspects of this case is the fact that, even after winning a decisive court victory against the drug consortium, the South African government refused to invoke the emergency clause that would have allowed it to import cheap drugs. In this way what had at first appeared to be a nationalist anti-imperialist struggle uniting the South African government and various NGOs against the global drug industry, was now dispersed across a heterogeneous, transnational theater of war—with the South African government pitted simultaneously against internal dissidents (most notably the Treatment Action Campaign, or TAC) and AIDS activists worldwide. It was only in 2003 and in the lead-up to national elections that the South African government made the seemingly cynical decision to reverse its policies, now promising to make antiretrovirals available by 2008.

How can we interpret the fatal, even genocidal denialism of the South African government, given President Thabo Mbeki's own acute awareness of the links between power and public health? Why did Mbeki, following in the steps of the U.S. government, refuse to declare AIDS a national health emergency? And how does this fit within the South African government's own politics of security? To fully answer these questions, it is necessary to understand the ways in which a neonationalist, identitarian politics such as Mbeki's is able to collude with the imperatives of neoliberal economic policy, even when it declares the most intransigent of *moral* wars against neoliberal imperialism. In South Africa the global politics of emergency feeds into the micropolitics of *exorcism*. Inevitably, it is the bodies of those who stand uneasily at the boundaries between the transnational labor market and the nation, the market and the family, threatening to escape both, that are caught in the cross fire of these antagonisms. In a context where the transmission of the HIV virus is predominantly heterosexual, it is the body of the undocumented migrant and the prostitute that come to signify the irresolvable tensions of neoliberal biopolitics.

TRIPS AND THE NEW PHARMACEUTICAL IMPERIALISM

For a piece of legislation with such far-reaching implications for the health and survival of whole populations, the TRIPs agreement was passed and signed with remarkably little controversy by the member states of the newly formed World Trade Organization in 1996.[6] TRIPs is the most comprehensive intellectual property (IP) agreement of the twentieth century and will no doubt shape the terms of global movements and struggles around drugs, epidemics, and health for many decades to come. It concerns two of the most promising new technologies of the twenty-first century—digital and biotechnologies—and provides a precedent for the privatization of "knowledge" industries of the future. The TRIPs agreement extends to all kinds of intellectual property rights, not just to patents but also copyright, trademarks, geographical indications, and industrial designs. It includes the most recent innovations in IP law—software codes and certain kinds of biological invention. It is effectively global in scope, since most countries are members or at least aspiring members of the WTO. Most important perhaps, the TRIPs agreement generalizes the exorbitant price demands of the United States' most profitable and politically influential industry: Big Pharma.

In the final stages of its development TRIPs was enthusiastically supported by the U.S., European, and Japanese delegations to the Uruguay Rounds of trade negotiations. Yet it had begun life as the brainchild of an extremely small group of private lobbyists, united together as the Intellectual Property Committee (IPC), all of whom were CEOs in the North American pharmaceutical, software, and entertainment industries.[7] In the years leading up to the final ratification of the accord, the IPC had engaged in a vigorous campaign to persuade the American business press, the public, and Congress that the globalization of stringent IP law was the answer to the economic ills of the United States. Their argument was simple—the ultimate cause of U.S. economic decline was to be found in decades of lost income from IPR on innovation. And in order to prove it, they produced an avalanche of statistics detailing the massive amounts of income that the United States had reputedly given away to the counterfeiters of the world. The chief culprits, they argued, were the newly industrializing economies of the developing world, such countries as India, parts of South America, and southeast Asia, which were beginning to catch up with the drug production facilities of their northern counterparts.

The tactics adopted by the IPC were audacious, considering that the United States itself had traditionally been quite reticent about overstringent IP legislation. But its arguments had the benefit of being self-fulfilling: the "innovation" industries they were seeking to promote did not yet benefit from worldwide patent protection, while some of its claims, such as the necessity of protecting software code sources or microbiological processes, were entirely new. Given the absence of relevant IP laws, to accuse the developing world of having *stolen* the rightful gains of the self-professed knowledge economies was tautological to say the least. Yet by the time the TRIPs agreement came into full effect (in 2005), the United States had designated intellectual property rights protection as a major security issue.[8] The IPC had not so much protected the U.S. innovation industries as created them ex nihilo.

Adopting a wider lens, the long history of negotiations leading up to the TRIPs agreement needs to be placed within the context of the shifting contours of imperialism, as the developmental ideal of the post–World War II era was shunted aside by the drastic reforms dictated by neoliberalism. Coauthors Peter Drahos and John Braithwaite (2002, 67) have argued that the campaign to globalize IPR on drugs was in fact initiated by U.S. pharmaceutical companies as a preemptive response to the emerging industrial powers of such postcolonial countries as India, which in the 1950s and 1960s were rapidly acquiring the capacity to produce generic drugs of their own, while not recognizing North American patent laws. Again, the seemingly disproportionate response of the U.S. pharmaceutical industry was not dictated by any straightforward calculus of loss. As Drahos and Braithwaite (ibid.) recount, companies such as Pfizer did in fact have drug production facilities in India, but profits from developing markets represented a derisory amount of their overall revenue. Arguably, what these companies feared most was the demoralizing effects that cheap drug production in the third world might have on the artificially inflated prices of the first. In other words the visible proof that something like "health for all" could be achieved at cut-rate prices threatened to bring down the price of drugs in the domestic U.S. market. The drug companies responded in kind, with an excessive display of force, by institutionalizing another kind of threat. Henceforth, the standard insider argument in favor of patent protection on drugs would go something like this: there is no innovation without patents; no promise of new medicines without the production of an attendant scarcity of health.

When the drug companies brought their case against South Africa, it was with the intention of enforcing this rule. The TRIPs compliance case was supposed to fix the international price of life, serving as a precedent for both the developing world and the internal U.S. market. When the case was dropped in April 2001, however, it was largely because it was beginning to have the reverse effect. Negative publicity was casting too harsh a light on the drug-pricing strategies of Big Pharma within the United States. Moreover, the campaign to oppose the drug consortium had brought together a dangerous coalition of activists; in the United States, where similar campaigns had been fought around drug-pricing strategies a decade earlier, AIDS activists were joined by African solidarity groups and students campaigning to oppose the drug patents held by their universities, while in South Africa, opponents included members of the African National Congress, advocates for sex worker rights, and gay activists (many of whom were members of the Treatment Action Campaign).

The inside story of TRIPs can be usefully read as a counterhistory of the bioinformation revolution. In particular, it militates against the idea that the so-called process of globalization, with its attendant shift toward knowledge and innovation economies, is embracing the entire world and is devoid of vectors of power or control. What it suggests, rather, is that the very value of knowledge—its surplus and its promise—is the result of a quite deliberate self-transformation of the U.S. economy and that of its allies, one that was pursued through the international organizations created in the post–World War II era, but with the ultimate effect of entirely redefining the landscape of world trade and imperialism. It also suggests that this moment of self-transformation was not so much a spontaneous move as a *response* to disruptions in the third world status quo.

THE ARITHMETIC OF LIFE AND DEATH: FINANCIALIZATION, DEBT, AND NEO-IMPERIALISM

The South African AIDS crisis is, however, much more than a story about medicine—or the lack thereof. As critics from the TAC activist Zackie Achmat to the South African President Thabo Mbeki have insistently pointed out, the unfolding of the HIV epidemic in sub-Saharan Africa and other developing nations is as much the result of the neoliberal policies imposed by the International Monetary Fund (IMF) and the World Bank over the past two decades, as the immediate, undeniable symptom of a rapidly mutating, maddeningly resistant virus. Indeed, one could go further along these lines and argue that the simultaneity of the North American–led biotech revolution and the troubling return of infectious disease of all kinds, in both the developing world and advanced capitalist centers, is symptomatic of the intrinsic contradictions of capitalism. The peculiarity of capitalism on this argument would lie in its tendency to create both an excess of promise and an excess of waste, or in Marx's words, a promissory surplus of life and an actual devastation of life in the present.

Any comprehensive analysis of the South African AIDS crisis then would need to go beyond the commercial strategies of the pharmaceutical industry and look at the shifting dynamics of imperialism over the 1980 and 1990s. The processes of financialization and debt creation that I examined in chapter 1 were key elements in the changing face of world imperialist relations over this period. The U.S. Treasury's ability to redefine itself as the evanescent focal point of world debt creation, I have argued, was of crucial importance in reigniting its economic growth and fueling the rise of such promise-laden sectors as the life science industries. Yet considered from the point of view of a country such as South Africa, it is the violence rather than the promise of the debt relationship that comes to the fore and defines the biopolitical consequences of the neoliberal counterrevolution.

The political economist Giovanni Arrighi (2002 and 2003) has identified the key moment in this counterrevolution as the period between 1979 and 1982, when the United States adopted a monetarist policy designed to reroute financial capital flows back into the domestic market. Up until this point U.S. power had been gradually eroding under the pressure of the rapidly expanding global markets in financial capital, markets that threatened to wrest economic control not only from the United States but from the nation-state per se. The monetarist turn in U.S. policy dramatically reversed this trend and simultaneously effected a seismic change in world power relations. From the world's principal creditor nation and source of liquidity in the postwar period, the United States transformed itself into the world's largest debtor nation, whose exorbitant power has subsequently been founded on the incongruous basis of a perpetually renewed and escalating budget deficit. The same process, however, has signified something diametrically different for sub-Saharan Africa, for when capital returned to flood the U.S. markets, it simultaneously exited the South, setting off the debt crisis of the 1980s and 1990s. It was using the leverage of these impossible debt obligations that the IMF and the World Bank were able to elicit "home grown" structural adjustment policies from the South African government.

Here, as elsewhere, neoliberal reforms included the liberalization of currency controls, the removal of subsidies and import barriers, a renewed focus on exports, and perhaps most important a sudden disinvestment of the state from any kind of public service provision (aside from military spending). The result has been a dramatic collapse of such basic public infrastructure as water provision, sanitation, and public health—all of which, quite apart from the risk of HIV/AIDS, has led to an increase in mortality rates from the most banal of infections. In his most recent work radical urbanist Mike Davis (2006) has identified the exodus of the indigent from rural to urban slum areas as the most visible sign of neoliberal imperialism. What is occurring here, he argues, is a strategy of systemic underdevelopment in which the compact of unequal exchange is replaced by pure neglect: in the urban slums of the postcolonial era, survival has become a game of intense *self*-exploitation, running the gamut from informal service work to biomedical labor (for example, the sale of organs or participation in clinical trials).

Other scholars, such as Saskia Sassen (2003) and Isabella Bakker (2003), have analyzed the structural synergy between debt servitude and the proliferation of various forms of highly mobile, feminized labor (affective, sexual, and domestic), pointing to the new "counter-geographies of survival" that have emerged in the wake of economic globalization. International debt, these theorists argue, has had a disproportionate effect on the realms of social reproduction and hence on the lives of women. The intensification of sex work in particular, both as a profession and as occasional work, is one instance in which the politics of debt servitude impacts in very immediate ways on the micropolitics of contagion, reinscribing the global politics of unequal exchange in the immediate bodily traffic of fluid, drugs, and money. All of these tendencies are visible in postapartheid South Africa—an increase in migrant flows from rural areas and other countries in southern Africa, a proportionate rise in the number of migrating women, most of them working as domestic, care, and sex workers—all in the context of rising social inequalities that are profoundly reshaping the binarized racial politics of the apartheid era.[9] How are we to interpret and respond to a situation such as this, where the macroeconomics of debt materializes in the most brutal of ways, drastically transforming the conditions of everyday life? Here I follow the work of Mike Davis (2006), Zygmunt Bauman (2004), and Adam Sitze (2004), who in their recent responses to neoliberalism have argued that a return to Marx's work on surplus population is long overdue.

Marx develops his fullest analysis of surplus population in his third volume of *Capital*, where he argues that there is a structural relationship between capital's moments of crisis, debt creation, and the periodic devaluation of human life. Marx's central thesis here is that capital's growth tendencies are internally hamstrung by an insoluble tension—that is, in order to maximize its own process of self-accumulation, capital needs to mobilize and promote the creative forces of human life, yet at the same time the imperatives of surplus-value extortion mean that it is constantly trying to undermine these very forces. The history of capitalism is littered with institutional responses to these countervailing tendencies—responses that have attempted to mediate between the reproduction of capital and the reproduction of human life. But such solutions, according to Marx, "are never more than momentary, violent solutions for the existing contradictions, violent eruptions that re-establish the disturbed balance for the time being" ([1894] 1981, 357).

Another way of saying this is that for Marx the mathematics of capitalist growth is fractal rather than dialectic—that is, it tends in the long run to deflect all mediations, descending into recurrent bouts of crisis in order to evade impending limits to growth. Marx identifies two primary tendencies at work here. On the one hand, there is the process of temporal revaluation, the movement by which capital withdraws from investment in the present and relocates in the futuristic realms of financial capital. The flight into financialization is the speculative response to crisis—a faith-driven attempt to relaunch the accumulation of surplus value at a higher level of returns, in the hope that production will at some point follow. This is the prophetic, promissory moment of capitalist restructuring, the kind of utopia that is celebrated in neoliberal theories of growth. On the other hand, the importance of Marx's analysis is to show that the promissory moment is necessarily accompanied by a simultaneous move to disinvest from, devaluate, and lay waste to whole sectors of unprofitable production. A prime target here, Marx suggests, is human life and its costs of reproduction, since "for all its stinginess, capitalist production is thoroughly wasteful with human material" (ibid., 180). In this way Marx's theory of capitalist transformation points to the dual nature of the debt relation: where capitalism promises on the one hand, it devalues on the other. The creation of surplus population, of a life not worth the costs of its own reproduction, is strictly contemporaneous with the capitalist promise of more abundant life. The violent crippling of growth coincides with the imperative of growth without limit.

In what sense is Marx's theory of surplus, debt, and growth still relevant to an understanding of capitalist relations today? The current period of crisis can only be fully understood as a response to—and deliberate demolition of—the Keynesian, nation-state centered regime of growth established in the post–World War II period. This was a model that placed particular importance on the possibility of establishing economic and social equilibrium through the simultaneous development of mass consumption and mass production. The philosophy of the welfare state was inseparable from the idea that the interests of production and reproduction could be mediated, that the growth of capital and the growth of populations could continue ad infinitum in conditions of general equilibrium.

Such promises weren't specific to the highly industrialized centers or advanced economies: the immediate post–World War II period also gave rise to the concept of development, with its division of the earth into first, second, and third worlds, over which would preside the newly created international economic institutions of the World Bank and the IMF. Together, what these institutions were supposed to promote was the worldwide standardization of growth; the idea (succinctly expressed in the notion of a standard of living) that cycles of life, whether of products or populations, were merging toward a common norm. Thus the primary thesis of development theory, as expounded by the economist Walt Rostow (1960), was that given the right spurs to industrial development, even the so-called third world would advance unerringly and inexorably through progressive stages of growth to someday attain the first world ideal of high mass consumption. But it is perhaps in the realm of public health that we find the most ambitious attempt to implement the welfarist ideal of normalized, nation-centered growth. After all, the standard props of mid-twentieth-century public health—quarantine, mass vaccination, and the theory of immunity—were quite literal expressions of the notion that cross-border

movements could be contained in a way that might mediate between the interests of economic growth and the life of the nation. With its faith in bodily sovereignty and the possibility of recognition, the mid-twentieth-century theory
of immunity concurs with the philosophy of the nation-state that threats are
always identifiable and peace, in principle, attainable.[10]

Herein lies the idiosyncrasy of neoliberalism as a political practice: where
welfare state nationalism saw the standardization of growth as a limit to be
attained in some utopian future, neoliberalism confronts it as a historical limit
to be *deflected*, no doubt because it is threatening to become all too attainable.
Faced with declining levels of productivity, the neoliberal response is to blame
the social state and its politics of national redistribution. The solution it proposes is simple: the state must divest itself of the burdens of social reproduction and redirect its energies toward the accumulation of capital beyond the
boundaries of the nation. In this way neoliberalism declares war against the
whole standardization of life that underlies the very idea of social-state nationalism. The truth it espouses—and this is the crude truth of capitalism, according to Marx—is that in the long run there is no mediation to be found between
the rhythms of production and consumption, no progressive transition from
third to second to first world that does not at some point enter into conflict
with the blunt need to increase the production of relative surplus value.

Neoliberalism announces the end of the mediations that were so central to
the growth strategies of the welfare state and developmental biopolitics: the
second world, the middle class, the family wage, the very notion of the *standard of living* all give way to extreme differences in the distribution of life chances.
Contrary to the philosophy of the social state, it teaches that the collective risks
gathered under the banner of the nation can no longer be (profitably) collectivized, normalized, or insured against. Henceforth, risk will have to be individ-
ualized while social mediations of all kinds will disappear. Given the conceptual
affinity between theories of biological and political immunity, it is surely not
incidental that over the same period there has been a profound rethinking of
dominant twentieth-century ideas about biological defense and resistance. In
recent theories of immunity, it is the stability of boundaries between the self
and the other, and hence the possibility of recognition, that is being put into
question. Today's immune systems, it would seem, are having trouble distinguishing between the self and the other (the autoimmune disease); are being
called on to marshal forces against the uncertain, unknowable threat (adap-

tive evolution); and are at risk of succumbing to states of permanent, overalert-ness (the allergic reaction).[11]

Neoliberalism thus formulates the whole problematic of danger in funda-mentally novel ways. What is at issue here is neither the state of exception, as Giorgio Agamben (1998) has argued, nor the state of immunity, as Roberto Esposito (2002) has claimed, both of which presume the exercise of sovereign power. The arguments developed by these theorists are appropriate for the specific instances of violence generated by nation-state politics, colonialism, and eugenics but fail to identify the accident form of neoliberal imperialism. The characteristic danger of the neoliberal era is the fractal or non-normalizable accident: the state of emergency understood as self-propagating event rather than sovereignty's constitutive outside. And nowhere is this politics of danger articulated more clearly than in the new humanitarian discourse, with its notion of the "complex emergency."

Below I attempt to understand how and why the new security discourse has come to figure infectious disease as a complex emergency, and what this might imply in terms of a politics of public health.

MILITARIZING CONTAGION: AIDS AS A GLOBAL SECURITY THREAT

The UN's decision to redefine AIDS as a security issue is symptomatic of a wider shift in international relations discourse, where the spheres of life and war are tending to merge. As such, the "global AIDS threat" pertains to a whole spec-trum of neologisms—from human to biological to ecological security—that were proposed in the late 1980s and 1990s as a way of reconceptualizing strategy in the post–Cold War era. The new security discourse has been turned toward widely different ends by diverse institutions and interests but finds its first explicit—and tellingly ambiguous—formulation in the United Nations Devel-opment Program's *Human Development Report 1994: New Dimensions of Human Secu-rity* (UNDP 1994) and the 1992 work by Boutros Boutros-Ghali, *An Agenda for Peace: Preventive Diplomacy, Peacemaking, and Peace-keeping.* It turns up slightly later in the lexicon of U.S. foreign policy, but even here covers a remarkably simi-lar set of concerns. If we take a look, for example, at the report entitled *U.S. Foreign Policy and the Four Horsemen of the Apocalypse: Humanitarian Relief in Com-plex Emergencies,* published in 1997, we find a comprehensive account of the new security concerns and their implications for American operations overseas (Natsios 1997). Each of these texts is interested in identifying the dangers that are likely to unfold from the breakdown of the postwar equilibrium between first, second (communist), and third worlds. Implicit in all of them is the idea that with the collapse of the Cold War stand-off and the consequent withdrawal of U.S. military aid from third world countries, the latter would be particularly prone to social implosion. It is in the developing world, and in particular sub-Saharan Africa, that the new security discourse locates the hotbed of emerging threats.

The literature on human security has tended to promote the idea that in the post–Cold War era infra-state conflict will take precedence over war between sovereign states. The new wars, it claims, are primarily of an internecine character—civil strife, ethnic violence, guerrilla rebellions, and coups d'état. Their prime targets are no longer the military installations of the enemy state but the life of the population itself; essential infrastructures such as water, transport, power, and food supplies, along with the matrix of social relations, at the intersection of race and gender, that constitute the nation (hence, it is claimed, a growing recourse to rape and ethnic cleansing as weapons of war). Far from seeking to reinforce the territorial integrity of the state, these conflicts involve situations in which the state turns against its own people or the people against themselves. Moreover, the new dangers no longer originate in the constitutional sphere or the arena of international relations, but rather they erupt from below and from within the fabric of social and biological reproduction, or else from above, from the biospheric or ecological level down.

Thus the most immediate threats to security are no longer those of a formal military nature. Rather, we are confronted with a plethora of everyday dangers, whose only common characteristic is their disruptive effect on social and biological life: "ecological damage, disruption of family and community life, greater intrusion into the lives and rights of the individual" (Boutros-Ghali 1992, 3). Continuing in the same vein, the UNDP's *Human Development Report 1994* (1994, 23) describes the province of human security as extending from such "chronic threats" as "hunger, disease and repression" to "sudden and hurtful disruptions in the patterns of daily life." The provenance of such risks, it is argued, is beside the point. Whether human-made or natural, intruding from the outside or within, what distinguishes them is simply their indifference to national, territorial boundaries.

Does the discourse of biological security represent a genuinely new development in political thought?[12] And more pertinent, to what extent does it break with the defining ideals of such postwar institutions as the UN, WHO, and the Declaration of Human Rights? An essential premise of these mid-twentieth-century conventions is the idea that life can be protected, if not absolutely then statistically, from the threat of both war and disease. In this way danger itself is pushed to the margins of the nation-state and understood as a threat impinging from the outside. The Declaration of Human Rights formalizes the welfare state ideal that the risks inherent in social reproduction can be collectivized and hence normalized within the space of the nation. The WHO charter, along with its contingent of quarantine measures, seeks to establish the idea that the vectors of disease are containable at the border. In the words of historian François Ewald (1986, 362, 397–99), the postwar security agenda implies truce and thus establishes a fundamental separation of spheres between military and civilian life, security and life (or more precisely, between military security and social security). This separation of spheres may be transgressed in practice but subsists in principle as the normative ideal regulating the welfare of nations.

In contrast, the new security discourse concludes that such boundaries are no longer tenable. Its message is simple: we can no longer assume that life can be quarantined against disease or protected, even statistically, from war. In fact, we can no longer even assume that the conditions of livable existence are containable within the space of the nation or liable to the sovereign decisions of the state. At a purely semantic level, then, the new security discourse ends up reabsorbing the whole sphere of biological, social reproduction, and sexual politics within the sphere of military concerns. It is intent on convincing us that at any rate life is always infused with danger. It therefore ends up merging what the Declaration of Human Rights sought to keep apart—military security and human welfare, the right to life and the conduct of war.

How should we respond to this shift in the scope and meaning of security? And how is it likely to translate on an operational level, both in international relations and the micropolitics of everyday life? The political consequences of such a shift are surely ambivalent. In his study of the changing landscape of war in Africa, for example, the international relations theorist Stefan Elbe (2002) points out that the securitization of AIDS makes perfect, though awful, sense in a context where increasing numbers of the military are infected and rape is regularly deployed as a weapon of war. Moreover, it could be argued that the new security discourse is simply reflective of changes in world imperialist relationships, faithfully rendering the all too tangible effects of neoliberal economic reform on the countries of the developing world. What more flagrant instance of a complex emergency, after all, than the IMF-imposed austerity program? And what more vivid account of a structural adjustment program than the following rendering of the complex humanitarian emergency, offered by a defense adviser to the U.S. government?

- First, the most visible characteristic [of the complex emergency], [is] civil conflict . . .
- Second, the authority of the national government deteriorates to such an extent that public services, to the degree that they ever existed, disappear, and political control over the country passes to regional centers of power . . .
- Third, mass population movements occur because internally displaced people and refugees want to escape conflict or search for food. Public health emergencies arise as dislocated civilians congregate in camps.
- Fourth, the economic system suffers massive dislocation resulting in hyperinflation and destruction of the currency, double-digit declines in the gross national product, depression-level unemployment, and the collapse of markets.
- Finally, these first four characteristics, sometimes exacerbated by drought, contribute to a general decline in food security. This frequently leads to severe malnutrition that, although at first localized, may degenerate into widespread starvation. (Natsios 1997, 7)

Yet while this excerpt offers a powerful descriptive account of the violences associated with market reform, it studiously diverts all attention from the actual policy agenda of neoliberal government. The human security literature, in other words, both avows and disavows the violence of neoliberalism. Although it seems to offer an indisputable testimony to the experience of neoliberalism—proliferating in threats, warnings, and dire prognoses—it simultaneously prohibits any serious reflection on its economic imperatives. Instead, it displaces insecurity and its affective equivalent—fear—from the structural violence of free-market economics to the transversal movements of people, viruses, and biological agents of all kinds. Displacement is precisely the mechanism that informs the U.S. government's positions on AIDS, security, and emergency. For while it has done everything to prevent the countries of the developing world from invoking a state-of-emergency clause (on the pretext that this would do violence to the economic rights of the drug industry), its own foreign policy discourse is moving to criminalize all kinds of transnational biological traffic— from the bodies of the undocumented, the poor, and the untreated to the virus itself.

This conflation of causes and effects is as evident in the international humanitarian discourse on AIDS and security as in U.S. foreign policy. In the UNDP's *Human Development Report 1994*, for example, there is already the suggestion that the biological threat is concurrent with the increase in illegal refugee flows from countries of the South to the North. In one extraordinary passage the UNDP report (1994, 24) states that human security "means responding to the threat of global poverty traveling across international borders in the form of drugs, HIV/AIDS, climate change, illegal migration and terrorism." Similarly, the National Intelligence Council's report (2000, 1) on global infectious disease explicitly mentions the rising number of legal and illegal immigrants from the "megacities" of the South as a significant vector for emerging, drug resistant infections.

As these documents make clear, the tendency to formulate AIDS as a global security threat needs to be understood in parallel with the securitization of migration itself. Not incidentally, the development of an AIDS security discourse in the 1990s occurred in a context where many countries were moving both to criminalize undocumented migrants and to adopt public health criteria as a means for restricting entry.[13] Moreover, the securitization of AIDS has a particular affinity with the public health fears that have traditionally accrued around prostitutes. Today the overriding concern with cross-border sex trafficking obscures the complex necessities and desires that lead women to participate in sex work and all too often lends itself to policies that end up criminalizing sex workers themselves (Saunders 2005). In a context where women dominate undocumented migrant flows and the global labor market is defined, at the lower end, by all kinds of highly mobile, feminized labor (sexual, domestic, affective), the circulating bodies of women are increasingly what is at stake in international concerns about cross-border traffic.[14] As a sexually transmitted disease, and one that has come to be associated with all kinds of illicit or deviant sexual relations, it is no surprise that the AIDS virus has come to carry much of the burden of these fears.

MBEKI: PUBLIC HEALTH, EXORCISM,
AND THE PARADOXES OF GLOBAL APARTHEID

It would be disingenuous, however, to attribute the whole blame for South Africa's AIDS crisis to the interests of globalizing capital without looking at the equally culpable policies of the Mbeki government. It is Mbeki after all who, for three years after the court case victory against the drug companies, delayed implementing the public health emergency clause that would have authorized him to import cheap drugs. In defense of his position, Mbeki has insistently cited the historical complicity between public health science, biomedicine, and the apartheid state, arguing that the received wisdom on AIDS prevention and public health is tainted with the legacy of institutional racism.

And without a doubt, the history of South African apartheid is intimately entwined with the rise of modern public health strategies and theories of contagion. South Africa's first segregationist law was the Public Health Act of 1883, which allowed local authorities to enforce quarantine and vaccination under general powers of emergency. It was followed in 1900 by the forcible removal of black Africans to segregated homelands, again on the pretext that they represented a public health risk.[15] The apartheid-era response to the rise of the AIDS epidemic in South Africa was entirely consistent with this heritage. Indeed, as pointed out by scholar Jeremy Youde (2005, 426), AIDS became the last-ditch argument of apartheid supporters, who employed "nearly identical rhetorical tools as used in the early 1900s to justify segregation along the lines of public health." The apartheid-era South African government was perhaps the first to openly militarize the threat of HIV, even going so far as to voice the now very common fear that the HIV-infected could conceivably become the agents of an internal terrorist attack.[16] And it was certainly the first to consider the uses of the HIV virus as an agent of biological warfare; among several initiatives, the South African government is reported to have developed a secret biological weapons program that was actively engaged in a project to transform the virus into a sterility-inducing agent to be spread among African women.[17] Along with this overwhelmingly military response to the HIV threat, however, the apartheid regime adopted a politics of selective neglect in the areas of public health, education, and prevention. The effects of such studied inaction, as noted by countless commentators, were overwhelmingly felt in the African homelands, and several conservative politicians openly voiced their

hope that the AIDS epidemic would effectively eradicate Africa's internal surplus population.

Mbeki has presented his singular response (or nonresponse) to the AIDS
crisis as a refusal to perpetuate this legacy. The struggle against AIDS in South
Africa has thus become a key element in his philosophy of black African re-
naissance, one that dictates a novel epistemology of political and biological
immunity. Like many other critics before him, Mbeki accuses the science of
AIDS and AIDS prevention of perpetuating systemic racism. His claims are
numerous and in many instances wholly justified: the wide currency given to
theories locating the birthplace of AIDS in Africa, as well as speculation about
the transmission of the disease from a primate to an African, are all products
of the racist imagination; the tools of Western public health are wholly unsuited
to dealing with the specificities of the disease in Africa, where heterosexual
transmission has always predominated; the imposition of the Western model
of AIDS prevention in South Africa amounts to the pursuit of apartheid-era
public health policies by other means.

Notoriously, however, Mbeki has also given public support to the ideas of
the so-called AIDS dissidents—scientists who take issue with the orthodox
nosology of the HIV virus. Referencing their work, he claims that the relationship
between the HIV virus and the AIDS syndrome has not been definitively
established and that the bundle of infections grouped under the name of AIDS
have in any case long been prevalent in the African community. Not only are
ARVs ineffective in treating AIDS, he argues; they are also toxic in and of themselves (perhaps even the cause of infection). In Mbeki's hands, moreover, the
specific quibbles of the AIDS dissidents meld into a much more extensive critique of the global political economy, in many ways similar to the one I have
outlined earlier in this chapter. The ultimate cause of AIDS, he claims, is poverty.
If it exists at all, the HIV virus is no more or less than the vector of global capital, the contagion by which the new imperialism insinuates itself into the newly
born African body politic. However, Mbeki makes the same claim for the supposed treatment for AIDS. If the big pharmaceutical companies are interested
in negotiating price deals for ARVs in South Africa, he argues, it is only because
they want to use clinical trials to propagate the virus of global capital in the
very guise of treatment. In this astonishing slippage of arguments, Mbeki ends
up postulating a causal equivalence between the HIV virus and the ARV drug,
pathogen and cure, identifying both with the vectors of global capital flows.

He advocates a position of super-immunity, one that would prevent all poisons and cures from entering the African nation.

All this sits uneasily with Mbeki's political record in other areas, though, where he has shown himself to be resolutely committed to the neoliberal agenda. As the political economist Patrick Bond (2001) has outlined in detail, Mbeki, like many fellow postcolonial neonationalists, is in fact much more attuned to the demands of deregulation than his rhetoric would suggest. Aside from being a close ally of the Clinton administration, his period in office has been characterized by assiduous obedience to a self-imposed debt-repayment schedule and major cutbacks in public spending on health, welfare, and education.

There is one area, however, in which the Mbeki government has spent with a vengeance: the defense forces. At the height of the AIDS crisis, and despite the absence of any credible threat to its territorial security, the South African government embarked on a huge spending spree to update and reinforce its conventional weapons capability (Van der Westhuizen 2005). And in its attempt to justify the decision, it employed a series of arguments that strikingly anticipate the now familiar rhetoric of the war on terror: the essential unpredictability of the new security threat, the fact that threats may arise from within as well as outside the state, and the consequent need for a preemptive strike force capability. From the Web site of the old South African Department of Defense, the *South African Defense Force* was thus able to warn that "the essential unpredictability of international relations, in which the unforeseen threat might materialize relatively quickly, together with internal potential for disorder, means that we need to retain a capability for defense against lawlessness—whether internal or external. . . . Although our peacetime defense force does not need to be as large as a wartime force, it cannot afford to become complacent" (quoted in Harris 2002).

In its 1998 *Defense Review* the South African Department of Defense specified that it was working within the framework of an expanded definition of security, one that would extend beyond the purely military sphere to cover the whole spectrum of so-called human security concerns: "The government has adopted a broad, holistic approach to security, recognizing the various non-military dimensions of security and the distinction between the security of the state and the security of the people. The greatest threats to the security of the South African people are socio-economic problems like poverty and unemployment, and the high level of crime and violence."[18]

Read against the slippages of Mbeki's own reflections on AIDS, in which the virus, the treatment, and the infected person all come to stand in for a generic, delocalized sense of threat, the obvious danger of this policy is that it might be used to defend the "African nation" against the "internal lawlessness" of its sick and dying underclasses. These security measures come at a time of rising xenophobia, as internal immigration from other African states threatens to exacerbate already rising social inequalities.[19] Mbeki's politics of preemptive super-immunity, in other words, is in danger of flipping over into something akin to an autoimmune reaction—a danger that might be considered symptomatic of neonationalist responses to neoliberalism.

Neonationalism is endemic to the new world order. It manifests in strikingly local guises, however, and in South Africa it is the language of witchcraft (or *muthi*) that has lent itself most forcefully to the expression of a pervasive sense of emergency. According to numerous accounts, the transition toward postapartheid has been accompanied by a general perception that witchcraft is on the rise.[20] This sense of everyday insecurity can no doubt be attributed in part to the intense physical danger that plagues life in South Africa. Rates of murder and rape are alarmingly high, while the postapartheid police in no way represent an institutional recourse against crime, for any sector of the population. But since the occult is presumed to work most effectively in silence, and surreptitiously, the fear of muthi has been most intense as a response to the AIDS epidemic (in South Africa a diagnosis of HIV/AIDS, if known at all, is most likely to remain a secret). In its symptoms and modes of transmission, HIV/AIDS seems to embody many of the properties commonly associated with witchcraft, where muthi can be made with "human body parts, intercourse (in all senses of the word) with mystical creatures, and the use of zombies" (Ashforth 2005, 41).

By displacement, moreover, the invisible, promiscuous forces of the occult have come to be associated with the flows of global capital, whose equally intangible movements seem to have such devastating effects on daily life.[21] The valences of muthi are strikingly ambivalent—it can work as both poison and cure, it can be wealth creating and wealth depriving, a means of illegitimate theft or a charm for protecting property (ibid., 41–44). It therefore lends itself perfectly to the structural ambivalence of a neonationalist politics such as Mbeki's, allowing him to designate the antiretroviral drug as a cure that might just as easily manifest as a poison, the invisible vector of global capital and thus an instrument of dispossession posing as a source of wealth.

The irony of Mbeki's politics is that in the name of the postcolonial African nation, he ends up replicating the mix of denialism, savage health care cutbacks, and stigmatization that characterized the Reagan era's response to AIDS. Perhaps more alarming, Mbeki's decision to escalate military expenditure in response to a hypothetical, unpredictable, internal threat is strikingly consistent with the most recent directions in U.S. policy on infectious disease and bioterrorism. What this suggests is a certain resonance between the experiences of urban racial and gender underclasses, whether these are located in post-apartheid South Africa or the ghettoes of North America. It also speaks of the collusive relationship that someone like Mbeki is able to sustain with the neo-liberal orthodoxy, even while he has appointed himself the spokesman of a certain form of populist, moral anti-imperialism.[22] In fact, the Mbeki government's investment in "human security" is closely attuned to rising levels of xenophobia in South Africa. Interestingly, this xenophobia cuts across the racial, class, and gender lines of the old apartheid state. It is directed against labor migrants from other countries in the region and brings together the threat of economic invasion with fear of crime and infection.[23]

The danger, as argued above, is that this particular form of anti-imperialism ends up incriminating those that lie at the intersection of transnational flows of labor and the residual institutions of the family and the nation—those bodies that seem to bear a disproportionate amount of "traffic." Traffic in blood, bodily fluids, viruses, clinical trials, money, and desire. In a context where all kinds of popular exorcisms are carried out on the bodies of those presumed infected, it is perhaps inevitable that the body of the prostitute (a body that so forcefully challenges the boundaries between the family and the market, commodification and self-exploitation) should be singled out as the site and source of contagion. Especially when rising numbers of sex workers in South Africa are migrants, either from rural areas or other countries in southern Africa. One response to this provocation, and one means of reasserting the proper borders of the nation, is to criminalize sex workers themselves. Certainly, this seems to be what is happening in South Africa right now, where the government has recently launched a "Moral Regeneration Campaign," calling for zero tolerance in relation to prostitutes even while it continues to actively promote its tourism industry.[24]

Again, the South African response to AIDS is strikingly resonant with Bush-era global health and foreign aid policy, which has also taken a distinctly moralist, indeed evangelical and fundamentalist, turn.[25] It is no coincidence that in South Africa too, the evangelical and Pentecostal churches have become some of the most prominent voices in the struggle against witchcraft.

George W. Bush's 2003 Global AIDS Plan is notable for its efforts both to open up new markets for the pharmaceutical industry and to relocate AIDS policy in the hands of foreign policy and security agencies rather than public health institutions. Along with this merging of public health, commercial, and military interests, he has also redirected AIDS prevention and treatment funds toward faith-based, abstinence-only initiatives. I will return to Bush's faith-based politics of life and its imperialist dimensions in the final chapter of this book. In the next chapter I discuss the militarization of infectious disease as it has unfolded within the United States. I look at the growing importance of the hypothetical biological threat in determining the agenda of both U.S. foreign policy and life science research. The history of TRIPs and its effects on the AIDS epidemic in sub-Saharan Africa represents the first step in the neoliberal militarization of public health. But this tendency is increasingly at work in American domestic politics. In this way Bush's politics of biological emergency brings humanitarian warfare back to the United States and turns its methods against its own surplus populations.

◻ 3 ◻

PREEMPTING EMERGENCE

The Biological Turn in the War on Terror

> It cannot be predicted or prevented; it can only be accommodated. . . . That is wild
> variation.
>
> —Benoit Mandelbrot, *The (Mis)Behavior of Markets:*
> *A Fractal View of Risk, Ruin, and Reward*

IN 2004, THREE YEARS AFTER THE SPORADIC AND STILL UNRESOLVED ANTHRAX
attacks that followed September 11, the Bush administration became the first
in U.S. history to implement a national defense strategy against biological threats.
In the same year the U.S. Congress also approved the largest ever funding project
for biodefense research, to be carried out over the following decade. The legislation, going under the name of Project BioShield, authorized $5.6 billion for
the purchase and stockpiling of vaccines and drugs against bioterrorist threats,
granted the government new authority to initiate research programs, and gave
it special dispensation to override drug regulations in the face of a national
emergency. At the same time a more secretive initiative was under way to establish four research centers for the testing of biological weapons defenses. The
United States, it seems, was preparing itself for an attack of epidemic proportions. But what exactly was it arming itself against?

In his public addresses on the topic President Bush seemed unsure whether
the deadliest threat would be more likely to emanate from a deliberate bioterrorist attack or from any one of the resurgent or drug-resistant infectious
diseases that now regularly afflict urban hospitals. Official documents declared
that infectious disease outbreak and bioterrorism should be treated as identical threats, in the absence of any sure means of distinguishing the two. The
confusion was further reflected in the allocation of resources. Much of the new
funding for biodefense went to institutions that had previously been engaged

in public health and infectious disease research, while the ailing biotech start-
ups of the genomics era were encouraged to reinvest their energies in the new
arena of military applications. For U.S. defense, it seems, the frontier between
warfare and public health, microbial life and bioterrorism, had become strategically indifferent. To be effective at all, the war on terror would need to arm
itself against the generic microbiological threat, from wherever it might emerge.

What are we to make of this "biological turn" in recent U.S. defense policy? And how should we respond to a security agenda that conflates public
health, biomedicine, and war under the sign of the emerging threat? This chapter argues that the growing interest in biological weapons is more than merely
tactical and goes hand in hand with a strategic redefinition of the tenets of U.S.
defense, one in which the doctrine of mutual deterrence is replaced by full-
spectrum dominance, counterproliferation, and preemption. As the United
States moves to integrate biological weapons into its defense arsenal, it is the
very nature of warfare, security, and the military threat that is being rethought,
at the interface between public health and security, the biotech industry, and
military research. In a context where the boundaries between biomedicine and
war are increasingly and quite deliberately blurred, it no longer makes sense to
restrict our critical horizons to the military sphere as it is traditionally defined.

This chapter seeks to unravel the diverse threads leading to the biological
turn of the war on terror, from the recent evolution of infectious disease research
to the volatile fortunes of the biotech industry and the internal transformations of U.S. defense. I am concerned not only with the newly formed institutional alliances between biomedicine and the military, but also (and more
important) with the conceptual exchanges that have taken place between the
two domains over the past few decades. In particular, I look at the complex
history of such notions as biological emergence, resistance, and preemption and
their crossovers into U.S. defense discourse. Only by exploring these exchanges
is it possible to understand and respond to the full spectrum of interventions
currently being pursued in the name of the war on terror.

GERMS AT WAR

One of the most eloquent legacies of twentieth-century public health can be
found in the idea that some kind of final "truce" would at some point be reached
between ourselves and infectious disease. Ever since the development of germ

theory in the late nineteenth century, modern biology had imagined humans and microbes to be engaged in a merciless war: a struggle for survival from which only one of us would emerge victorious. Only after World War II, however, would public health institutions have the confidence to declare that the war was almost over; that infectious disease would be conquered once and for all, sequestered, quarantined, eliminated even, first in the "developed" world and later in the "developing" world, through the classic public health strategies of quarantine and immunization as well as the massive use of the new generation of antibiotics and vaccines. As late as 1978, the United Nations issued an accord predicting that even the poorest of nations would undergo an "epidemiological transition" before the year 2000, transporting us into a new era in which the chronic diseases of old age would prevail over infection.

Ironically, this was precisely the period in which infectious disease made a dramatic comeback. At a time when public health expenditure was being heavily cut back in the name of welfare reduction, and microbiology had long been relegated to the margins of the life sciences, new infectious diseases were on the rise again while old diseases were reemerging in new, more virulent forms. This in any case was the view from the richest countries, which had long considered themselves immune from the plagues still raging "over there." In the year 2000 the World Health Organization officially announced in its *Report on Infectious Diseases* that the truce was over: the return of infectious diseases worldwide represented a deadlier threat than war; we had been caught off guard; the microbes had been preparing an underground counterresistance just when we thought we were finally safe.

The militaristic language of classical germ theory made a spectacular return to public health discourse, but this was warfare of a different kind, one that unsettled the reassuring dogmas of the quarantine state. Pathogenic microorganisms were proliferating from within and without. Friends were turning against us. The immunological self was misrecognizing itself (the autoimmune disease). Our most promising cures (antibiotics) were provoking counterresistances at an alarming rate. The apparent triumph of biomedicine was generating its own blowback effects (due, for example, to the overuse of antibiotics in the "developed" world and their underuse in the "developing" world). Diseases that had long been considered chronic or genetic were suddenly revealing an unsuspected link to latent infections (P. Ewald 2002). New pathogens were crossing borders that were supposed to be impenetrable, including frontiers between species (such as mad cow and Creutzfeldt-Jakob disease). Contagions were hitching a ride on the vectors of free trade (the deregulated blood market that enabled the contaminated blood scandals to happen; the complex cross-border movement of food implicated in mad cow disease), perhaps even on the mobile vectors involved in the production of transgenic crops and therapeutics.

The same era witnessed something of a conceptual revolution in microbiology. The new microbiology tells us that our relation to microbial life is one of inescapable coevolution. We are literally born of ancient alliances between bacteria and our own cells; microbes are inside us, in our history, but are also implicated in the continuing evolution of all forms of life on earth. Biologists are discovering the biospheric dimensions of microbial life (the notion of a common evolution linking plants, animals, and microbes with the geology of the earth and the composition of the atmosphere) and claiming that emerging infectious diseases are indissolubly linked with climate change. In the words of biologists Lynn Margulis and Dorion Sagan (1997, 94) the environment "is regulated by life for life," and the common vector linking all these life forms and responsible for maintaining a breathable atmosphere is provided by bacterial evolution.

At the same time recent research is throwing new light on the specific processes of bacterial evolution, suggesting that bacteria evolve through highly accelerated processes of horizontal communication rather than chance mutation and selective pressures. It has been known, since the late 1950s, that bacteria are able to exchange sequences of DNA, often between unrelated species, through a general process of horizontal transfection.[1] Only recently has the full extent of this mobility become apparent: under certain conditions mobile sequences of bacterial DNA jump across species, genuses, and kingdoms; once integrated into a new genome, these sequences are able to mutate and recombine; the bacterial genome itself is highly fluid, capable of mutating under stress and accelerating its own mutation rate (Ho 1999, 168–200). While many leading infectious disease specialists continue to see microbial resistance as a form of (highly accelerated) Darwinian evolution (Lederberg, Shope, and Oaks 1992), a growing body of new research is suggesting that bacteria do not even have to wait around for random mutation to confer resistance; they can share it among themselves. The new microbiology is discovering that for bacteria resistance is literally contagious (Levy and Novick 1986; Ho 1999, 178–79).

These new insights into microbial resistance have important ramifications for our understanding of genetic engineering technologies. What molecular biology shared in common with the political philosophy of twentieth-century public health was the belief that the future evolution of life could be predicted, controlled, and (at worst) reverse-engineered on the basis of localized interventions. This shared utopia is coming under increasing scrutiny, however, as recent research points to the possible links between the reemergence of infectious disease and the use of recombinant DNA technologies. The production of "transgenic" life forms, after all, hitches a ride on the same vectors of communication that are responsible for resistance—viruses, transposons (mobile genetic elements), and plasmids (extrachromosomal genetic elements)—while these vectors are routinely modified to render them even more prone to circulate and recombine. As the full extent of horizontal transfer comes to light, biologists are beginning to suggest that we cannot mobilize these vectors of communication without provoking and even accelerating the emergence of all kinds of counterresistance.[2]

EMERGENCE REEMERGING

The microbiologist René Dubos was the first to coin the term "emergence" as a way of describing the temporality of biological evolution. By "emergence" he understood not the gradual accumulation of local mutations, but the relentless, sometimes catastrophic upheaval of entire coevolving ecologies; sudden field transitions that could never be predicted in linear terms from a single mutation (Dubos [1959] 1987, 33). Writing at a time when the "health transition" was official public health doctrine, Dubos dismissed the idea that infectious disease could ever be eliminated, let alone stabilized. There can be no final equilibrium in the battle against germs, he argued, because there is no assignable limit to the coevolution of resistance and counterproliferation, emergence, and counteremergence. In Dubos's work the concept of microbial "resistance" is divested of its association with the pathological: resistance is merely another word for emergence, and there is no end to it; its future evolution is unforeseeable from within the present.

Dubos is scathing in his criticism of the strategic vision of mid-twentieth-century public health, but what he offers in response is not so much a pacifist manifesto, as an alternative vision of warfare and a counterphilosophy of disease. If we are at war, Dubos contends, it is against an enemy that cannot be sequestered—a threat that is not containable within the boundaries of species life, is both inside and out, necessary for our survival yet prone to turn against us, and capable of reinventing itself in response to our "cures." Dubos's theater of war presupposes a coimplication of human, bacterial, and viral existence; a mutual immersion in the conditions of each other's evolution. It is inevitable—he argues—that our most violent efforts to secure ourselves against contagion will be met with counterresistances of all kinds. Microbial life will overcome our defenses, and yet we can never be sure when and how it will happen: "At some unpredictable time and in some unforeseeable manner nature will strike back" (ibid., 267).

If we are to follow Dubos, the relentless nature of coevolving emergence irresistibly engages us, despite ourselves, in a form of permanent warfare, a guerrilla counterresistance without foreseeable end, against a threat whose precise "when" and "how" we can only speculate on. Such an elusive vision of warfare might seem to preclude any effective strategic response—yet it is at this level that public health must organize, according to Dubos. If humans are to survive the inevitable "counterstrike" from microbial life, he argues, we need to prepare for the unexpected; learn to counter the unknowable, the virtual, the emergent. The new science of life, Dubos writes, must cultivate "an alertness to the advent of the unpredictable," a responsiveness to the threat that is merely felt or apprehended (ibid., 271). We must become capable of responding to the emergent, long before it has actualized in a form we can locate or even recognize. Life is a gamble, Dubos contends—a kind of speculative warfare (ibid., 267). And war is necessarily preemptive, as much an attempt to resist the countercontagion as a creative reinvention of the conditions of human existence, beyond whatever actual limits we might have adapted to in the present.

At the time he was writing (the 1950s), Dubos could not have been more at odds with the reigning public health orthodoxy. Three decades later, however, his counterphilosophy of disease seems to have been taken up into the mainstream of microbiology. The continuing evolution of infectious disease is inevitable, microbiologists now tell us. There can be no final conquest of infectious disease, although nothing will allow us to predict when and where the next pandemic will emerge. As biologists Joshua Lederberg, Robert Shope, and Stanley Oaks (1992, 32) have written: "It is unrealistic to expect that humankind will win a complete victory over the multitude of existing microbial diseases, or over those that will emerge in the future." They continue: "Although it is impossible to predict their individual emergence in time and place, we can be confident that new microbial diseases will emerge" (ibid., 1992, 32). The new public health discourse calls our attention to *emerging* and *reemerging* infectious disease; old pathogens that have resurfaced in new, more virulent or resistant forms; existing pathogens that have infected humans for the first time; or entirely new creations. It defines infectious disease as *emerging* and *emergent*—not incidentally, but *in essence*. What public health policy needs to mobilize against, the new microbiology argues, is no longer the singular disease with its specific etiology, but *emergence itself*, whatever form it takes, whenever and wherever it happens to actualize.

More ambiguously, the new discourse on emerging infectious disease seems also to have struck a chord with U.S. foreign policy and international relations theorists, who over the same period were busy at work enumerating the new and "emerging threats" that would define the post–Cold War era of warfare. Under the banner of the new intelligence agenda certain defense theorists (often with the uncritical support of NGOs and humanitarian organizations) were arguing that the scope of security should be extended beyond the conventional military sphere to include *life itself* (Johnson and Snyder 2001, 215–18). At issue here is the securitization of human life (hence the altogether strange concept of humanitarian warfare), but increasingly U.S. defense discourse is wanting to push further and incorporate *the whole of* life, from the micro- to the eco-systemic level, within its strategic vision.[3]

One of the most prominent advocates of the concept of microbiological security has long claimed that "emerging infectious disease . . . poses a clear threat to national security" and that U.S. defense should develop a common strategy for confronting both emerging and drug-resistant disease and bioterrorism (Chyba 1998, 5). In case this might seem to represent an extreme position, it is worth noting that in 2000, a CIA report classified emerging "global infectious disease" as a nonconventional security threat comparable to the new terrorism (National Intelligence Council 2000). In 2002 the U.S. Congress passed the Public Health Security and Bioterrorism Preparedness and Response Act outlining the same emergency response procedures for bioterrorist attacks and epidemics (U.S. Congress 2002). More recently, the Pentagon has published a report exhorting the U.S. government not only to wake up to the impending threat of climate change (assumed now to be closely related to the resurgence of infectious disease) but to treat it as a national security threat (Schwartz and Randal 2003). The future evolution of life, it warned, would be defined by permanent warfare.

Importantly, the securitization of emergence has occurred in parallel with the neoliberal demolition of public health—a process whose effects have been felt in everything from access to community-level care to vaccination programs and health insurance. In this way the new security discourse takes up Dubos's philosophy of microbial war while divesting it of all its profound insights into the relational practice of living with germs. For Dubos the essential point is that microbial life, with its inherent mutability, evolves in close synergy with ecologies of public health. Biological phenomena such as drug resistance, while inevitable in the long run, can therefore be intensified or countered by the way we administer and distribute medicines. Hence the tendency of the rich to overuse antibiotics and of the poor to underuse them are both factors in the acceleration of drug resistance.

Furthermore, our institutional preparedness for the surprise biological event can be more or less hampered by the organization of such basic public health services as garbage collection, sewerage, access to water, and free medical care. It is difficult to assess to what extent the neoliberal dismantling of public health is responsible for accelerating the emergence of new diseases, although it is surely a contributing factor. However, one incontestable result of Reagan-era public health cutbacks has been a general loss of preparedness in relation to the emergent event. In 1987 the Institute of Medicine warned that the Reagan-era demolition of public health had left the United States structurally incapable of dealing with even the most familiar of infections.[4] It seems that nonpreparedness in the face of the surprise event is a distinguishing feature of the neoliberal politics of risk more generally. Paradoxically, while neoliberalism insists on the utter unpredictability of the biological threat, it also establishes it as inevitable and pervasive. The event will have taken place, although we can never be sure when or how. And although we are exhorted to *feel* prepared, it leaves us constitutively unprepared for even the mildest of surprises.

BIOSPHERIC RISK: COUNTERING THE EMERGENT

Throughout the 1980s a new understanding of risk turned up simultaneously in the language of insurance institutions, capital markets, and environmental politics. This was the concept of the "catastrophe risk."[5] The catastrophe event was discovered in the guise of the global environmental disaster. From nuclear winter, global warming, and ozone depletion to emerging disease and food-borne, transgenic, and biomedical epidemics, the "catastrophe risk" has come to designate a technological accident of biospheric proportions, operating simultaneously at the microscopic and the pandemic levels. More recently, it has merged with the concept of the "complex humanitarian disaster"—a state of social breakdown that defies the simple predictive strategies of the Cold War period.[6] What is at issue here, according to the historian François Ewald (1993), is a fundamentally new calculus of the accident. Unlike the punctual accident of classical risk theory, the catastrophe cannot be insured against. The changes it threatens to introduce are irreparable, "not only because their scale exceeds the capabilities of any indemnity-providing organization" but also because their long-term effects "affect life and its reproduction," life and its vectors of communication (ibid., 223). Inscribing itself in the ecological conditions of life on earth, the catastrophe event is disturbingly both destructive and "creative."

If the catastrophe event is routinely presented as something of a paradox of risk management, it is because it confounds the traditional framework of rational decision making. Classical risk theory presumes that we can predict the likelihood of a future event, at least in statistical terms. The longer our time scale and the wider our field of vision, the more accurate our predictions will be. If we feel that we are unable to calculate the probability of an event, we can always wait until more information becomes available before making a decision. Prediction founds the possibility of prevention. At worst, classical risk theory reassures us that if the accident does occur, we will have been able to insure against it. Catastrophe risk, however, denies us the luxury of preparation. When and if it happens, it will be by surprise, abruptly, and on a scale that overwhelms all efforts at damage control. What we are dealing with here is not so much the singular accident, as the accident amplified across a whole event field, a phase transition that may emerge without warning, instantaneously and irreversibly transforming the conditions of life on earth.

To make things worse, the nature of these events is such that we can never be sure how far gone we already are. Disaster is incubating. We may be on the verge without realizing it. It may already be too late to slow down, to reverse the process, to restore some kind of (relative) equilibrium. If the catastrophe befalls us, it is from a future without chronological continuity with the past.

Although we might suspect something is wrong with the world (look at those freaky weather patterns, those locust plagues, melting ice caps, and emerging pandemics, for example), no mass of information will help us pinpoint the precise when, where, and how of the coming havoc. We can only speculate.

What we do know, however, is that if such an event were realized, its consequences would be catastrophic, irreversible, and of incalculable cost. As the environmental risk theorist Stephen Haller (2002, 93) has put it, "we cannot afford not to decide," and yet catastrophe risk places us in the uncomfortable position of having to take drastic and immediate action in the face of an inescapably elusive, uncertain threat, decisions that may in turn generate their own incalculable dangers. "My concern," he continues, "is about the general problem of what to do in cases where we are asked to take action meant to avoid catastrophe before we have compelling evidence of the likelihood of the catastrophe" (ibid., xii). "We must face squarely the problem of making momentous decisions under uncertainty" (ibid., 87).

Here François Ewald has identified the defining predicament of the neoliberal politics of security. The catastrophe event, he writes, confronts us with a danger we "can only imagine, suspect, presume or fear"; a danger we "can apprehend without being able to assess" (F. Ewald 2002, 286). In this sense the new discourse of catastrophe risk establishes our affective relation to the future as the only available basis for decision making, even while it recognizes the inherently speculative nature of this enterprise. What it provokes is not so much fear (of an identifiable threat) as a state of alertness, without foreseeable end. It exhorts us to respond to what we suspect without being able to discern; to prepare for the emergent, long before we can predict how and when it will be actualized; to counter the unknowable, before it is even realized. In short, the very concept of the catastrophe event seems to suggest that our only possible response to the emergent crisis (of whatever kind—biomedical, environmental, economic) is one of speculative *preemption*. Again, in the words of Haller (2002, 14), writing, it should be noted, before the transformation of preemption into official U.S. strategic doctrine: "Some global hazards might, in their very nature, be such that they cannot be prevented *unless pre-emptive action is taken immediately*—that is, before we have evidence sufficient to convince ourselves of the reality of the threat. Unless we act now on *uncertain* claims, *catastrophic* and *irreversible* results might unfold beyond human control."[7]

At this point it is important to distinguish between two postures of preemption that have begun to make their place in international politics. On the one hand, the so-called precautionary principle represents a counteractive response to emergent catastrophe risk: in the face of an uncertain future it advises us to halt all further development of a technology suspected of harboring some kind of latent risk factor. The biologist Mae-Wan Ho (1999, 168) has cautioned that "we may already be experiencing the prelude to a nightmare of uncontrollable, untreatable epidemics of infectious diseases" and that on the basis of this suspicion "we must call a halt [to genetic engineering] now, there is no time to lose."

Remarkably, the principle of precaution has been formalized in such international accords as the Kyoto protocol and the legislation of certain EU countries, where it introduces the novel legal principle of a duty to undertake collective preventive action in the face of the unforeseeable. The text of a French law, approved in 1995, perfectly captures the philosophy of precaution when it states that "the absence of certainty, taking into account the state of scientific and technical knowledge, must not postpone the adoption of effective and proportionate measures to prevent serious and irreversible damage to the environment" (cited in F. Ewald 2002, 283). Acting in the name of a generalized suspicion, the precautionary principle is perhaps less progressive than it might at first appear. It finds its political counterpart in neoliberal social policies that dismantle the buffers of the welfare state only to criminalize the slightest acts of deviance. Zero tolerance is the sociological face of environmental precaution.

On the other hand, the concept of preemption is increasingly being brandished as a justification for aggressive counterproliferation, particularly in the United States. This is most obviously the case of the U.S. government's new doctrine of military preemption. But the move to preemption was already visible in the United States' changing position on environmental, biotechnological, and biospheric risk. Under George W. Bush the United States withdrew from the Kyoto protocol and a new UN agreement to enforce the Biological and Toxins Weapons Convention (BTWC) of 1972 (although it should be noted that Bill Clinton had already initiated bioweapons research that flouted the nonproliferation accord on germ warfare). What the United States is beginning to formulate here is a legal right to aggressive counterproliferation, where the point is no longer to halt innovation on the mere suspicion of its incalculable effects but precisely *to mobilize innovation in order to preempt its potential fall-out.*

In the economic domain one very practical application of catastrophe risk has been the invention of new speculative instruments such as cat bonds, which since the mid-1990s have allowed reinsurers to hedge for natural and technological disasters on the capital markets. Catastrophe bonds covering natural and aerospace catastrophes are now regularly traded, but proposals have been made to issue titles for everything from acts of terrorism to climate change and genetic accidents. The usefulness of the catastrophe bond, in place of the more cautious asset investments of the past, is again commonly attributed to the very nature of the catastrophe event, which is declared to be uninsurable, at the limits even of the calculable (Chichilnisky and Heal 1999). As one industry report put it, the potential for accidents associated with the new biotechnologies demands that we "think the unthinkable and quantify the unquantifiable" (Swiss Re 1998). The catastrophe bond resolves the apparent dilemma by transforming uncertainty itself into a tradable event, protected by a legally binding contract. In the process it invents a form of property right that seeks to capture the speculative biological future at its most unpredictable—literally, before it has even emerged.

It is all of these aspects of the catastrophe event—economic, biospheric, and military—that come together in the new strategic discourse on bioterrorism.

EMERGING THREATS

When the Nixon administration renounced its biological weapons program in 1969, it was because germ warfare seemed to offer none of the advantages of the nuclear bomb or chemical weapons. In their submissions to a Senate inquiry into biological warfare, U.S. defense advisers argued that germ warfare was naturally resistant to the strategic aims of mutual deterrence and should be abandoned: biological agents were unpredictable in their effects, responsive to uncertain climatic and environmental conditions, indifferent to national borders, and prone to backfire on those who used them, making it difficult to defend the boundaries between the civilian and the military spheres, friend and enemy, here and over there (Novick and Shulman 1990, 103; Wright 1990, 39–40). Not only was biowarfare unworkable within the strategic framework of mutual deterrence, they claimed; it threatened to undermine the very "balance of powers" on which this doctrine was predicated.

Several of Nixon's advisers warned that the dissemination of germ warfare would lend itself to nonstate resistance movements, democratizing the use of weapons of mass destruction in a way that would permanently undermine the strategic advantage of both the United States and the Soviet Union (Wright 1990, 40). What bioweapons threatened to propagate was not only a specific pathogen but another mode of warfare altogether. Beyond their immediate and deadly rivalry, the superpower states thus shared a common interest in preventing the emergence of nonsovereign enemies. For all of these reasons, it seems, the United States had no qualms in unilaterally giving up its offensive bioweapons program, whatever the USSR chose to do. In 1972 the Biological and Toxins Weapons Convention, banning the use and possession of biological weapons, was signed in London, Moscow, and Washington, D.C.[8]

Three decades later, biowarfare has moved back from the margins to the center of U.S. defense policy, while the doctrine of mutual deterrence has given way to the war on terror, full-spectrum dominance, and preemptive strikes. In 2001, Bush inaugurated his presidency by withdrawing the United States from a new UN effort to enforce the BTWC of 1972. And following the anthrax attacks of 2001, Bush called on the U.S. Congress to approve a massive decade-long funding scheme for "defensive" bioweapons research. The era of biological nonproliferation was officially over.

Like many of Bush's more spectacular military maneuvers, the turn toward a strategy of biological counterproliferation was already prefigured in the institutional reform process that was the so-called revolution in military affairs (RMA). Initiated in the early 1990s, the RMA was never anything more than an attempt to *simulate* the hypothetical future of warfare, and yet it also set forth a number of tacit prescriptions for the strategic reorganization of U.S. defense (many of which were carried out under Clinton). Informing this literature was the certainty that the solutions of the Cold War era were no longer capable of shoring up the hegemonic position of a superpower state such as the United States.[9] The era of state-centric, bipolar conflict had established a certain kind of equilibrium—the shared risk aversion of mutual deterrence. With the collapse of the former Soviet Union, however, it could no longer be assumed that the capacity for mass destruction would remain the sole prerogative of the superpower state.

The RMA literature predicted that twenty-first-century warfare would be dominated by terrorism, but of a different type than the more familiar kinds of state-sponsored terrorism. The new terrorism might be funded by one or several states (think of the relationship between the Saudi Arabian elite and neofundamentalist Islam), and yet in its modes of violence, dissemination, and recruitment it would operate outside of territorial boundaries. Rather than classify these "emerging threats" according to their national, political, or ideological alliances, the RMA literature highlighted their common indifference to the state-centered logistics of the Cold War period. With no territory to defend, the new enemy could not be contained within the affective limits of mutual deterrence (mutual fear as a source of risk aversion). Nor could it be countered with traditional models of prediction, risk assessment, and decision making. In the words of defense specialist Anthony Cordesman (2001, 421), "there is no 'standard distribution curve' of past events that can be used to predict the future" of terrorist attack. Terror is by definition "uncertain," "emerging," and pandemic. Hence its "catastrophism," according to Clinton's defense advisers.[10]

At the same time the RMA anticipated that the rise of catastrophic terrorism would rehabilitate biowarfare as a viable military option. The nuclear and chemical arms of the Cold War period, underwritten by the massive industrial infrastructure of the superpower states, might not become completely obsolete but would be progressively marginalized by information and especially biological warfare. Revelations about the bioweapons program of the former Soviet Union and the exodus of its scientists into Iraq, followed by Iraq's own admission of a smaller program, fed into media-channeled fears that the United States had dangerously neglected this "weapon of the poor." The Clinton administration pointed to various abortive attempts at anthrax attacks by cult groups in the United States and Japan as a sign that the new warfare would be bioterrorist, while bioweapons experts warned that genetic engineering provided new opportunities for the creation of novel, highly virulent pathogens (Block 1999; Miller, Engelberg, and Broad 2001; Fraser and Dando 2001). The idea that biological agents would be the weapon of the future thus hardened into official public discourse.

But how do we assess this overwhelming, highly mediatized conviction on the part of the U.S. government that the future of warfare will be biological, given that the actual instances of bioterrorist attack in the United States remain rare, underwhelming, and (in the case of the 2001 anthrax attacks) of dubious origin? Whatever the likelihood of these future scenarios, the sudden preoccupation with biowarfare needs to be understood, above all, as an effect of the deliberate *self*-transformation of U.S. defense, a revolution in military affairs that in any case threatens to blur the difference between real and imagined threat. U.S. strategy has moved full circle since the Cold War. Where once the point was to stave off the emergence of minoritarian, nonstate guerrilla movements (or at least to recuperate them within the struggle against communism), the United States now aims to prevent the reemergence of a Soviet-style superpower state.[11] In line with the strategic vision of the RMA the United States is restyling itself as an emergent guerrilla resistance movement on a worldwide scale (albeit one supported by massive state-deficit funding), transforming war into a process of permanent neoliberal counterrevolution.[12]

As a consequence, the doctrine of mutual deterrence has been demoted as the organizing principle of U.S. defense. Under Clinton it was tentatively replaced by the concept of counterproliferation—a move that was criticized as a first step toward preemptive warfare.[13] The Bush administration has gone further and merged counterproliferation with full-spectrum dominance and preemption to formulate a pervasive, future-oriented space-time of military responsiveness. It is in this particular strategic context that the United States has come to affirm the importance of biological weapons research. U.S. defense advisers and bioweapons experts now claim that the very traits that made biological weapons so useless for the Cold War superpowers are precisely what might recommend them to the new generation of terrorists (Chyba 1998 and 2000). More pertinently, it seems clear that U.S. defense is incorporating bioweapons research of an ostensibly defensive nature into its own long-term restructuring of military affairs. At stake here is much more than a tactical reorganization of military R & D, weapons stockpiling, and funding. The potential usefulness of biological warfare, as envisaged by U.S. defense, is both strategic and affective: or rather strategically affective because, as noted by specialists in the burgeoning field of terrorism psychology, biological weapons "are especially effective at causing terror" (Hall et al. 2003, 139). With their ability to spread without detection, to incubate and produce delayed effects, biological agents are capable of transforming emergence itself into the ultimate military threat. In the early twenty-first century, it would seem, bioterrorism is becoming the focal threat of U.S. defense policy—the virtual, characteristically emergent event around which it is reorganizing its whole vision of warfare.

PREEMPTION

As various commentators have pointed out, preemption is not a new concept in international law. Traditionally, however, the right of preemption authorized a state to counterstrike when it had warning or visible evidence of an *imminent* attack. The U.S. National Security Strategy of September 2002 outlined a radically new doctrine of war that specifically legitimates the use of preemptive action against a threat that is not so much imminent as *emergent*, a threat whose actual occurrence remains irreducibly speculative, impossible to locate or predict.[14] Unlike the reliable Cold War opponent, Bush warned, the new terrorist networks and rogue states are oblivious to the persuasive force of mutual deterrence. Their movements are incalculable, uncertain in time and place, of indeterminable cost—and this, we are told, is precisely why the United States cannot afford to wait:

> The greater the threat, the greater is the risk of inaction—and the more compelling the case for taking anticipatory action to defend ourselves, *even if uncertainty remains as to the time and place of the enemy's attack.* To forestall or prevent such hostile acts by our adversaries, the United States will, if necessary, act preemptively; . . .
>
> America will act against *emerging threats before they are fully formed.*[15] (National Security Strategy 2002, 15, 4)

The preemptive strike has been decried as a departure from all existing principles of legitimate warfare. But it would be better understood (and countered) on its own terms, as a radically new formulation of law, one that founds the legitimate use of violence on "our" collective apprehension of the future, however uncertain, rather than the predictive calculus of risk. In this sense the concept of preemption has more in common with the principle of precaution, itself increasingly at work in international *environmental* law, than any prior doctrine of warfare. Both preemption and precaution endow our suspicions, fears, and panics with an active force of law. Both insist on our absolute, uninsurable exposure to an uncertain future, our coimplication in events that recognize no sovereign boundaries. But whereas the precautionary principle advises on a course of absolute intolerance to the future, the doctrine of preemptive warfare assumes that the only way to survive the future is to become immersed in its conditions of emergence, to the point of actualizing it ourselves.

Preemption transforms our generalized alertness into a real mobilizing force, compelling us to become the uncertain future we are most in thrall to. As a mode of anticipation, it is *future invocative* rather than predictive or representative, since the future it calls forth is effectively generated de novo out of our collective apprehensiveness. What the U.S. Security Strategy of 2002 wants to affirm—by force of law—is that the mobilizing condition of warfare can only be speculative. The Department of Defense thus stipulates that institutional and strategic transformation of the military "should be thought of as a process, not an end state. Hence, *there is no foreseeable point in the future* when the Secretary of Defense will be able to declare that the transformation of the Department has been completed. Instead, the transformation process will continue indefinitely. Those responsible for defense transformation must *anticipate the future and wherever possible help to create it*" (Office of Force Transformation 2004, 2).[16]

Since it was first elevated to an official doctrine of U.S. defense in 2002, the concept of preemption has traveled far outside its original context and is increasingly at work in U.S. policy on emerging environmental and health crises, ranging from global warming to infectious disease. In 2002, shortly after Bush's national security report was released, the editorial of one foreign policy journal suggested that the new doctrine of preemption should be extended to climate change:

> . . . by pushing for preemptive military action in the name of national self-defense, the United States has forced a new post-Westphalian definition of the limits of sovereignty when facing the new cross-border threats of the twenty-first century. And those threats include not only terrorism, but climate change as well. . . . Like future terrorist actions, we can't be absolutely certain what will happen, but all the signs are there. . . . Rather than wait until it is too late—when floods, droughts, rising sea levels, melted glaciers and new diseases abound—why not take the wise course and preempt that possibility by acting now. . . . Whether the Bush administration comes to this view or not, its new preemptive doctrine has already galvanized the international community, inadvertently providing a rule book and a logic for multilateral action on other cross-border threats, including climate change. (Gardels 2002, 2–3)

This journalist was by no means alone in his vision of the military future. In late 2003 the Pentagon published a report (see Schwartz and Randal 2003) on the potential consequences of abrupt climate change for U.S. security.[17] The authors of the report outline the now familiar dilemma of catastrophe risk:

although the risk of climate change is inherently "uncertain"—Will it happen? Is it happening already? How severe will the effects be? Are we on the verge of some irreversible phase transition?—its consequences are "potentially dire" and therefore necessitate urgent action (ibid., 3). The point, according to the report, is not only to "think the unthinkable," but more important to actively preempt the emerging catastrophe through what the authors refer to as adaptive strategies. In particular, they suggest that the United States should explore "geo-engineering" options designed to transform the earth's climatic conditions by unleashing various active gases into the atmosphere. What the Pentagon is proposing, then, is a "solution" that is both speculative and biotechnological (in the widest sense of the term). It recommends that we intervene in the conditions of emergence of the future before it gets a chance to befall us; that we make an attempt to unleash transformative events on a biospheric scale before we get dragged away by nature's own acts of emergence.

In the meantime the Defense Advanced Research Projects Agency (DARPA), the Pentagon's center for funding cutting-edge military technology, is working on a similar response to the problems of emerging infectious disease and bioterrorism (Miller, Engelberg, and Broad 2001, 306–7). One of DARPA's current projects includes the creation of biological sensors—living cells on chips or three-dimensional cell matrices—that respond to both known and *previously uncharacterized* agents to give a warning sign of attack.[18] But DARPA's research is not limited to advanced detection technologies; it is also engaged in the development of drugs that are similarly responsive to the unknown. Using the technique of DNA shuffling (hailed as the second generation of genetic engineering because of its highly accelerated capacity for randomly recombining whole segments of genomes), DARPA is attempting not only to perfect our defenses against existing threats but more ambitiously to create antibiotics and vaccines against infectious diseases *that have not yet even emerged.*

Molecular geneticists associated with this research have appropriately referred to DARPA's experiments with the DNA shuffling method as a form of anticipatory evolution (Bacher, Reiss, and Ellington 2002). While this research is being carried out under the banner of biodefense, DARPA finds itself in the paradoxical situation of having first to create novel infectious agents or more virulent forms of existing pathogens in order to then engineer a cure. Blurring the difference between defensive and offensive research, innovation and preemption, the Pentagon seems to have decided that aggressive counterproliferation is the only possible defense against the uncertain biological future. This is a "solution" without reprieve—if the emergence of biological resistance is inexhaustible, DARPA's preemptive war against evolving infectious disease and bioterror can only be of indefinite duration.[19]

Already, biologists are warning that the massive new biodefense research institution being built at Fort Detrick, in Frederick, Maryland, looks like it is preparing for both offensive and defensive bioweapons research (Leitenberg, Leonard, and Spertzel 2004). In any case the very nature of the bioweapon makes it almost impossible to disentangle the two.

BRINGING HUMANITARIAN WARFARE BACK HOME: THE WAR 'ON TERROR AND DISASTER RESPONSE

The vicissitudes of bioweapons research, however, are only one part of the story behind the U.S. government's shifting policies on infectious disease. Another essential element can be found in the recent history of humanitarian intervention, where the boundaries between the realms of war and civil life—social, biological, and environmental reproduction—have become increasingly difficult to sustain. Throughout the 1990s the discourse on humanitarian intervention effectively challenged the rules of legitimate warfare, introducing the novel idea that states, under the authority of the UN Security Council, had the right and even the duty to intervene militarily in the internal affairs of other sovereign states in order to pursue humanitarian or disaster relief. A military intervention, for example, might be justified on the basis of a general breakdown in social and urban infrastructure, whatever the cause—ethnic conflict, economic devastation, infectious disease, or environmental catastrophe. International relations experts were now openly anticipating that the major refugee flows of the twenty-first century would be caused by environmental crisis and conflicts over resource scarcity. In the words of George Bush Sr., the whole field of humanitarian relief and disaster response consequently found itself invested with "new-found military and diplomatic implications," which the United States, in particular, did not hesitate to avail itself of (Bush 1997, xiii–xiv).

More recently, the United States has begun to turn the logic of humanitarian intervention inward, directing its energies toward the militarization of a whole spectrum of potential homeland emergencies—from terrorist attacks to epidemics and freak weather events. Two U.S. policy documents are particularly illuminating on this development. If we look, for example, at *U.S. Foreign Policy and the Four Horsemen of the Apocalypse: Humanitarian Relief in Complex Emergencies*, a report published by the Center for Strategic and International Studies in 1997 and prefaced by George Bush Sr., we find a comprehensive account of the scope and strategies of humanitarian warfare (Natsios 1997). The new security threats, it stipulates, are defined by the fact that they imperil human, biological, and even biospheric existence rather than the formal political institutions of the state. A second document, *The Global Threat of New and Reemerging Infectious Diseases: Reconciling U.S. National Security and Public Health Policy*, published by the RAND Corporation in 2003, furnishes a remarkably similar argument in favor of the expansion of security concerns (Brower and Chalk 2003). This document, however, is concerned with America's *domestic* affairs. Purporting to "reconcile U.S. national security with public health policy," it recommends that the government adopt the methods of humanitarian intervention as a way of responding to domestic "biological security" threats running the gamut from emerging infectious disease to deliberate incidents of germ warfare. The implicit conclusion is that public health crises—indeed natural disasters of all kinds— require the same kind of full-spectrum military response as deliberate acts of terror.

A key turning point in this direction was the Bush administration's 2003 decision to incorporate the Federal Emergency Management Agency (FEMA) within the Department of Homeland Security, turning it overnight from a civil emergency response agency into an antiterrorist organization. The decision, along with the appointment of a wholly inexperienced director, has been attributed to Bush's serial incompetence. Yet it no doubt also reflects a more far-reaching transformation in policy direction. The reinvention of FEMA, for example, occurred at the very moment that Bush announced the BioShield program and then proceeded to reorganize infectious disease research and former public health institutions under the aegis of biodefense. At work here is a similar logic to the one deployed a decade earlier in defense of humanitarian warfare. It is a logic that on the one hand points to a general breakdown in the structures of public health, sanitation, and emergency response (attesting to but not questioning the effects of neoliberalism), and on the other hand declares that militarization is the only solution.

Moreover, it appears to culminate in the idea that the new security services are best left in the hands of the private sector. *The Four Horsemen of the Apoca-*

lypse, for example, not only stipulates that response to natural disasters should be militarized, but also that public funding of warfare and its attendant humanitarian missions should be channeled into all kinds of private initiatives, contracted out, deregulated, and subject to the demands of profitability (Natsios 1997, 33–75). This of course is nothing new in the area of humanitarian intervention overseas, where private service providers, from nondenominational NGOs to faith-based, abstinence-only initiatives, are increasingly the norm.[20] But it does represent a significant new development in the response to American *domestic* crises. And in case we needed some proof of the kinds of intervention such policies might engender, we have only to look at the disastrous unfolding of the administration's response to Hurricane Katrina or at Bush's stated plans for dealing with a possible outbreak of avian flu.[21]

The first point to be made is that Hurricane Katrina was as much a "natural" disaster as the revelation of years of government neglect. If proof were needed, Katrina made manifest what it means to deliberately neglect or undo the preventive arsenal of urban infrastructure, sanitation, and grassroots public health services that make up a kind of invisible barrier against the catastrophic accident. As countless commentators have detailed, the disaster defense mechanisms of New Orleans had been allowed to disintegrate in the years preceding Katrina, even when public officials warned of the possible consequences. All of this meant that when an accident happened, and accidents always happen, it was bound to be catastrophic. The second point is that the promised disaster relief either did not happen at all or happened selectively or happened too late, so that television audiences around the world could participate in the surreal experience of witnessing in real time the nonarrival of rapid-response forces.

When it did happen, the belated "rescue effort" had all the strangeness of a humanitarian intervention waged on U.S. soil. With the finesse of American soldiers shooting to kill in Somalia or dropping fatal food packages over Afghanistan, FEMA is reported to have arrived on the scene to distribute vials of the eagerly awaited anthrax vaccine. And when the National Guard turned up, it was seemingly with the intention of protecting the city from those who were stranded there: troops were deployed around the white, gentrified, and commercial districts; bridges and exits cordoned off; people remaining in the city (mostly poor and African-American) were prevented at gunpoint from collecting water and food, if not detained. Survivors were declared to be "refugees" (later rebaptized as "evacuees") and dispatched to one of several temporary housing camps, many of which look like becoming semipermanent institutions for detaining the poor. In the meantime, as in any other humanitarian intervention, disaster response and military personnel were followed by a motley crew of NGOs, private charities, and faith-based initiatives—all of whom were presumably supposed to take in charge the unenviable task of posthurricane relief. But it is the plans for posthurricane reconstruction that are most revealing about the growing indistinction between U.S. foreign policy, internal disaster response, and urban regeneration: with many of the key reconstruction contracts going to Halliburton, the principle contractor in postwar Iraq, it would appear that the war on terror is being pursued against America's own racial and class minorities.

All of this might be interpreted as a simple lapse in vigilance, the combined result of overstretched forces, political neglect, and plain incompetence—except that Bush has gone even further in his suggested plans for confronting outbreaks of infectious disease such as a potential avian flu epidemic. Pointing to the logistical role of the National Guard in New Orleans, Bush has concluded that the armed forces should be granted an even wider margin of maneuver in future "catastrophe-like" events, while key figures in the administration have gone so far as to propose a suspension of the Posse Comitatus Act of 1878 (an act that does not prohibit, but seriously limits, the use of armed forces in civil law enforcement). The implications, as pointed out by one conservative commentator, are enormous. On a purely formal level the suspension of Posse Comitatus would mean that the declaration of a national health emergency would automatically translate into a state of martial law, while on an operational level, it would fully transfer the task of responding to an epidemic into the hands of the Department of Defense.[22]

In his public announcements on the matter Bush brings together the established motifs of unpredictability and probable catastrophic consequences with the requisite admission of resignation in the face of the inevitable (it is by now too late to mass-produce vaccines, stock up on sufficient antivirals, or rehabilitate the public health service). All of which, in a now familiar unfolding of logic, adds up to the necessity for preventive (read "preemptive") action. Specifically, Bush has suggested that the Department of Defense should be authorized to forcibly isolate, evacuate, and quarantine the first line of infected people in any pandemic.

ECONOMIES OF EMERGENCE

In the mid-1990s the official rate of U.S. productivity growth suddenly took off in the statistics after a long twenty-five-year slump, seeming to confirm that the "information revolution" was indeed beginning to bear fruit. This sudden burst of exuberance was hailed as the sign of an emerging postindustrial revolution, whose two cutting-edge sectors (biotechnology and information technology) would relaunch the U.S. economy into a golden era of indefinite growth. As venture capital flooded into the digital and biotechnologies, it seemed that speculation itself had become the driving force behind unprecedented levels of innovation, allowing whole industries to be financed on the mere hope of future profits. What was at stake here, even according to the most skeptical of observers, was much more than an irrational bubble or the delirious financialization of the economy (Brenner 2002). Far from representing a final abstraction of the virtual from the tangible world of bodies, the rise of venture capitalism institutionalized a model of economic growth in which *production itself* was made to hinge on the vagaries of stock-market investment. This could not have been more evident than in the biotech sector, where the most material of productions—the experimental regeneration of life itself—became intimately infused with the virtual temporality of speculation.

The political theorist Christian Marazzi has described the venture capital model of accumulation as an economy of emergence, where the so-called fundamentals of production are replaced by the traditional affective skills of the professional speculator—the ability to sense and respond to crowd movements before they take hold; to initiate new product lines before a market exists for them; to promote belief, euphoria, or panic in the face of an event that has not yet materialized. Marazzi (2002, 48–49) writes: "Everyday productivity is increasingly determined by the capacity to respond in unforeseen and unforeseeable situations, *emergent situations*, those situations that obviate any kind of programming and posit *occasionality* as central."[23]

During the late 1990s whole sectors of the economy were held aloft on a wave of media-induced expectation—expectation of profit, in the first place, but also a kind of collective faith in the soon-to-be realized possibilities of the new information and life science technologies. At a time when most biotech companies had yet to develop a marketable product, let alone make a profit, capital investment in the new technologies was sustained by the hope that the

Human Genome Project (HUGO) and genomics in general were about to deliver an unheard-of revolution in health care, an era of designer drugs and precision-targeted interventions into the germ line. In March 2000, though, the venture capital frenzy of the late 1990s came to a fittingly millennial end when the dot-com stocks collapsed, followed later in the same year by the mass anticapitalist protests in Seattle.[24] It was in this atmosphere of impending political and economic crisis, announcing the decline of the neoliberal triumphalism of the Clinton era, that Bush came to power. In retrospect, it seems clear that the war on terror was as much a political response to the downturn of the new economy as to the terrorist attacks of September 11. Bush's answer to the technophilic optimism of the Clinton era was an equally megalomaniac plan for indefinite war, encompassing the whole globe within his strategic vision.

For a while venture capital continued to invest in the life sciences, with the lingering hope that the promised new economic growth would at least become tangible here. But when HUGO and other genome sequences were published, there was a sudden sobering consensus that the life sciences would need to move into the "post-genomic" era before the anticipated medical breakthroughs could be realized. In 2003 the fortunes of the biotech sector slumped to an all-time low, and at this point the U.S. government came to the rescue with a massive plan to fund "biodefense" research for the following ten years. The plan included generous incentives for drug development that seemed as much designed to overcome the time lags of commercialization as to counter the threat of bioterrorism. New biodefense legislation made sure that any "national health emergency" would become the perfect occasion for pushing through a drug without clinical trial.[25] Biotech would live again, but this time federal funding of life science research would be tagged to the new strategic vision of the Bush administration. The long-neglected domains of public health and infectious disease research would be rehabilitated and merged with biodefense while venture capital investment would again be courted, but this time on the pretext of permanent war rather than permanent growth.

The difference between Clinton's neoliberalism and Bush's neoconservatism needs to be qualified then: both economies mobilize speculative affect, attuning it to the emergence of the unpredictable. What has changed is the affective valence of "our" relation to the future—from euphoria to panic to fear, or rather alertness (that is, a state of fear without foreseeable end). Where the celebrants of the new economic growth reassured investors that there was no end to innovation, holding hope aloft with a constant barrage of short-lived promises, the neoconservatives want to convince us that there is no end to danger, that the war against terror can only be indefinite in time and scale.[26] In the aftermath of September 11, permanent warfare has become the new driving force behind U.S. economic growth, feeding off its own ineptitude as it generates a seemingly inexhaustible demand for security services of all kinds. Within this new configuration of powers, the life sciences have been promoted to a commanding position. The Bush administration has achieved something the theorists of Clinton's new intelligence agenda only ever dreamed of—the actual institutional conflation of security and public health research, military strategy, environmental politics, and the innovation economy.[27]

What is being articulated here is a profoundly new strategic agenda where war is no longer waged in the defense of the state (the Schmittian philosophy of sovereign war) or even human life (humanitarian warfare; the human as bare life, according to Giorgio Agamben [1998]), but rather in *the name of life in its biospheric dimension, incorporating meteorology, epidemiology, and the evolution of all forms of life, from the microbe up.* The extension of preemptive warfare to include the sphere of environmental and biopolitics conflates the eternaliza-tion of war with the evolution of life on earth—as if permanent war were sim-ply *a fact of life,* with no other end than its own crisis-driven perpetuation. As Dick Cheney has said: "It may never end. At least not in our lifetime" (quoted in Woodward 2001).

Inevitably, such a delirious prognosis on the future of warfare demands that we rethink the shape of a possible antiwar politics. Perhaps, given the recent nature of the events analyzed in this chapter, the problematic of resistance can be most forcefully posed in the interrogative form. What becomes of an anti-war politics when the sphere of military action infiltrates the "grey areas" of everyday life, contaminating our "quality of life" at the most elemental level (Brower and Chalk 2003)? In what sense is it even possible now to claim a right to "life," social security, public health—the peculiarly vital rights of the welfare state—without falling into the trap of legitimating permanent warfare? And how do we counter a politics that turns the possibility of ecological crisis into a tradable catastrophe risk on the capital markets?

One response to these questions has been to redefine security in human, biological, or even biospheric terms, as if this were the only way to salvage something of the vitalist politics of the welfare state. But such a strategy falls straight into the hands of the new intelligence agenda, with its manic desire to revitalize and expand the scope of legitimate security interventions. Rather than plead for a security politics with a human face, then, a more promising vector of resistance lies in the attempt to undermine the nexus between military security, the politics of life, and new forms of speculative capitalization. In the face of a politics that prefers to work in the speculative mode, what is called for is something like a creative sabotage of the future; a pragmatics of preemptive resistance capable of actualizing the future outside of the policeable boundaries of property right. And in the face of a politics that all too often adopts a posture of resignation in the face of the biospheric catastrophe, it is imperative that we do not give in to the sense of the inevitable. Neoliberalism has a vested interested in selective fatalism. Perhaps then the task of a counterpolitics is twofold: wherever possible, all efforts should be made to undermine the foregone conclusion, and when all else fails, the aim should be to reroute the catastrophe toward more interesting ends (catastrophes often become the occasion for renewing and creating countercommunities).

This is an abstract formula for resistance that applies to such diverse questions as the capitalization of health and old age insurance; biological patents of all kinds; and even the commercialization of the "elements," from privatized water to tradable pollution rights and environmental catastrophe bonds. Such a formula could describe any number of recent conflicts around the neoliberal politics of life, from the court case opposing AIDS activists to pharmaceutical companies in South Africa; to the revival of popular pharmacologies in the face of the depredations of a global drug market; and to projects in open source biology initiated by scientists across the life sciences, to name but a few. What is new about the current context, however, is the creeping militarization of these sites of biopolitical tension. The domains of life that neoliberalism has sought to incorporate into commercial and trade law over the past two decades are now being forcibly recruited into an expansive politics of military security. Increasingly, then, any counterpolitics of health, ecology, and life will need to engage with the pervasive reach of the war on terror; to contest, in other words, the growing collusion between neoliberalism's politics of life and the imposition of a permanent state of warfare.

And it will need to do so while resisting the temptation of the neoreligious, survivalist response to catastrophe. It is quite striking that both evangelicals and Salafist Islamists responded to Hurricane Katrina by interpreting it as an act of divine revenge, occasioned not so much by American foreign policy as by sexual decadence. In the eyes of the neofundamentalists the uninsurable catastrophe risk—commonly referred to as an act of God in insurance parlance—is precisely that: an act of God calling for a faith-based initiative of humanitarian, urban, and spiritual regeneration.[28] Neofundamentalism is the toxic by-product of neoliberal catastrophism, and while the anticapitalist left has concentrated on the latter, it has missed the micropolitical level, where survivalism, witch hunts, and piousness translate the abstract economic event into a divine thunderbolt, visited upon the infidels of this world. It seems to me that the one cannot be countered without the other.

Intermezzo

THE FIRST THREE CHAPTERS OF THIS BOOK WERE CONCERNED WITH THE disciplines of molecular biology, microbiology, and infectious disease research. There I was primarily interested in biotechnologies that mobilize the productive capacities of microbial life—recombinant DNA, bioremediation, and germ warfare. The chapters moved from a consideration of the most promissory, utopian impulses animating the biotech revolution, to an analysis of the structural violences inherent in contemporary forms of pharmaceutical imperialism, and finally to the literal convergence of life production and warfare in the global war on terror. However, I do not wish to suggest any finality to this sequence of ideas—the chapters could just as easily be arranged in the opposite direction.

In the second half of the book I turn from the biotechnical arts of recombination to the emerging sciences of regenerative medicine. This is an area that draws on a very different genealogy of life science disciplines—from embryology and developmental biology to oncology and reproductive medicine. I begin with a consideration of the technical novelty of tissue engineering and its difference from both organ transplantation technologies and Fordist modes of mass production. What is at issue here, I suggest, is a thorough rethinking of the possibilities of bodily transformation, one that needs to be read in parallel with the space-time imperatives of post-Fordist production techniques. I then move on to a consideration of the interfaces between stem cell science and reproductive medicine, and attempt to delineate the new forms of labor and accumulation that are crystallizing around the production of embryonic life. Finally, I turn to the most insistent and pernicious of life politics today—the right-to-life movement—with its manic desire to reestablish the fundamentals of life and (re)production, even in the face of the most speculative of life science technologies.

Here I come full circle, since the right-to-life philosophy is not entirely alien to the neoliberal utopias of perpetual growth and earthly regeneration that I examined in chapter 1. Or rather, the right-to-life movement embodies the contrarian impulses of capitalism, bringing together the promissory futures of neoliberalism with a violent reimposition of limits, fundamentals, and values. What it gives voice to is perhaps nothing more than the contemporary form of capitalist contradiction. But if this is true, what needs to be investigated is why the tensions of contemporary capitalism are concerned, above all, with the production and generation of new life. What might this mean for a politics that seeks to counter both the neoliberal and the neofundamentalist tendencies of contemporary power relations?

CONTORTIONS

Tissue Engineering and the Topological Body

Up until now, the question of the relationship between inert matter and life has above all focused on the problem of fabricating living matter from inert matter: the properties of life were located in the chemical composition of living substances. . . . However, a hiatus remains between the production of the substances used by life and the production of the living being itself: in order to affirm that one is approximating life, one would have to be capable of producing the topology of the living being, its particular type of space and the relationship it establishes between an interior and exterior milieu. The bodies of organic chemistry are not topologically distinct from the usual physical and energetic relations. However, the topological condition is perhaps primordial in the living being as such. There is no evidence that we can adequately conceptualize the living being within the framework of Euclidian relations.

—Gilbert Simondon, *L'individu et sa genèse physico-biologique*

T HE FIELD OF REGENERATIVE MEDICINE, WHICH COMBINES STEM CELL SCIENCE and tissue engineering, has been hailed as a second-generation model of earlier biomedical technologies, such as prosthetics and organ transplantation. It has also been associated with a return to "mechanistic" or "architectural" theories of biology in which the engineering of forces and relations (stress, tension, compressibility, cell-surface interactions) predominates over the semiotics of code, message, and signal.

Emerging out of a heritage of reconstructive medicine and in vitro cell and tissue culture, the aim of tissue engineering (TE) is to reconstruct three-dimensional living organs and tissues in vitro, from the cellular level up, to then transplant them back into the patient's body. Unlike reconstructive medicine TE does not simply transfer tissues through microsurgery techniques; it also works to modulate their morphogenesis, in vitro and in vivo. The sources from which living cells can be cultured are multiple—so far, biologists have used the cells of aborted fetuses, frozen embryos, the abandoned foreskins of circumcised children, as well as other discarded tissues, but the most ambitious proposals seek to use therapeutic cloning as a source of autologous (self-to-self) donation of embryonic tissue. Once sourced, these cells are then cultured and multiplied in a three-dimensional form, often with the aid of some kind of scaffold, which may be made out of natural, synthetic, or biodegradable material (or some mixture of these). Thus far, biologists have had most success in developing structural substitutes such as skin (the first commercially available TE product), bone, cartilage, and heart valves, but experiments are also under way to construct more complex metabolic substitutes to compensate for liver and pancreas failure. Another area of study, which I am less concerned with here, involves the direct implantation of stem cells (neural, hematopoietic, and islet) into the body.

The first experiments in tissue engineering date back to the early 1990s, but the field as a whole really took off as a result of advances in stem cell biology in the late 1990s. The successful culturing of pluripotent human embryonic stem (ES) cell lines, along with the discovery that adult stem cells were more ubiquitous and more plastic than had previously been thought, led to a deeper understanding of the body's possibilities of transformation. The reconstruction of organs and tissues, it was thought, could be envisaged not only as a process of in vitro morphogenesis, but more ambitiously as one of reproducible embryogenesis.[1] TE has been credited with the potential to overcome the intractable problems associated with organ transplantation and prosthetics—immune reactions, the scarcity of transplantable organs, the limited life span and wear-and-tear of medical implants in the body. The fact that TE works with the regenerative possibilities of the body (its ability to recreate itself) would mean that organ scarcity and use-by dates would no longer be an issue. The use of autologous transfers, or even the manipulation of cell-surface interactions, it is predicted, will displace the problem of immunogenicity. In this sense TE is hailed as an upgraded version of these earlier biomedical technologies.

But is there an essential continuity between TE and the organ technologies of the mid-twentieth century? Are they working with the same concept of animation, of bodily (re)generation and transformability? In this chapter I argue that the techniques of reproducible morphogenesis exploited in TE differ in fundamental respects from the biomedical paradigm of organ transplantation and prosthetics. These differences are first of all of a disciplinary nature: biochemistry, immunology, cryopreservation, surgery, and mechanical engineering were all essential to the development of organ transplantation and medical devices, while tissue engineering makes use of the latter techniques but is more closely related to the areas of developmental biology, morphology, experimental embryology, and regeneration studies. The field of tissue engineering, as its name suggests, has fostered all kinds of cross-disciplinary alliances between biologists, materials scientists, and engineers (chemical, mechanical, and even electrical), but the specific problems it encounters have also gone hand in hand with a rethinking of traditional mechanistic models of engineering.

Most important, these differences can be understood as a function of the mathematical theory of *transformations*, which distinguishes between various geometries and the kinds of movement they allow. While early formulations of prosthetics and organ transplant technologies draw on the rigid, metric transformations of kinematics (an appellation that significantly combines a reference to mechanical motion and the study of images in motion, or cinematography), the continuous modulations of force deployed in TE are more evocative of the mathematics of *topological* space. Here the rigid bodies of metric space are plunged back into a field in which discontinuous forms are continuously "morphable" into each other. The point-to-point movements of pregiven bodies give way to the *morphogenesis* of form as process. In order to argue this point, I look in detail at the various methods of biological modulation currently being developed in TE (such as the use of bioreactors, computer-assisted modeling, parametric variation) as well as more general developments in topological computer design. I am also interested in the privileged historical relationship between topology and embryology.

While recent cultural and design theorists (Lynn 1999; Cache 1995) have drawn on the resources of topology to develop a philosophy of nonmetric architectural and bodily space, I suggest in this chapter that developments in regenerative medicine demand that we think through the conditions not only of nonmetric space but also of *nonmetric time*. By opening up the technical possibility of reproducing the morphogenesis of the body outside of metric and genealogical time, TE confronts us with the enigmatic phenomenon of "permanent embryogenesis."

ARTS OF THE ORGAN:
PROSTHETICS AND ORGAN TRANSPLANTATION

Prosthetics and organ transplantation both underwent a surge of development after World War II. With the war-driven invention of new materials and later advances in electronics and software, this period witnessed the first large-scale industrial production of prosthetic substitutes for missing organs and bodily functions. In general, these devices took over a mechanical, optical, acoustic, or electric function of the body. The most successful among them include artificial joints, plastic lens implants, hearing aids, pacemakers, and cardiovascular devices. Other devices, such as the dialysis machine or heart-lung machine, attempt to substitute for the physiological function of the body from the outside. More recent developments have seen electronics and software integrated with traditional prosthetics to create bionic or robotic devices.

It was also in the post–World War II period that the essential components of organ transplantation—including cryopreservation, immunology, blood and organ banking—were developed to a workable level, even though the first successful attempts at transplanting organs date back to the beginning of the century. The first kidney transplants were performed in the early 1950s, followed by heart and liver transplants in the 1960s. And with the arrival of powerful immunosuppressant drugs in the 1980s, organ transplantation became a more or less routine method of bodily repair. Emerging over the same period of time, organ transplantation and prosthetics as procedures of mass biomedicine appear to share little in common besides the goal of bodily reconstruction. Transplantation, after all, is an art of the living biological organ, while prosthetics exploits the techniques and materials of the industrial machine. But despite their difference in materiality, both technologies share a mechanistic vision of animation, one that assumes the fundamental equivalence of the organ and the machine.

THE MECHANICS OF ANIMATION:
THE PHILOSOPHY OF ORGAN TECHNOLOGIES

In his classic 1992 study on the "Machine and the Organ," the philosopher of science Georges Canguilhem argued that the mechanistic theory of biology only came into its own with the invention of automata—machines that are able to fuel themselves from an internal source of energy, independently of the muscular force of a human being or animal. Up until then, he observed, the machine's capacity for motion could be considered separately from its capacity for self-preservation. It was only when the mechanical device was combined with the motor, mechanics with energetics, that the idea of machinic animation acquired a certain scientific plausibility (Canguilhem 1992, 104–6).

One of the most thorough explorations of the mechanistic theory of life can be found in the work of physiologist and inventor Étienne-Jules Marey, whose various biomedical, photographic, and cinematographic devices inspired many of the new technologies of the early twentieth century, including prosthetics and (arguably) organ transplantation (Rabinbach 1992, 90–110). Writing in the 1860s and 1870s, Marey developed the idea that the theory of energy conservation could be used to explain all phenomena, biological or otherwise, and therefore invalidated the vitalist belief that life followed laws of its own. Modern science had succeeded in creating "animate motors"; therefore in principle it should also be capable of engineering mechanical life. Underlying these different manifestations of force was a common principle of production or labor—muscles perform mechanical work through the combustion of nutrients, just as the steam engine produces energy by burning carbon. The physicochemical labor of the body was measurable in the same terms as that of heat. There was thus a fundamental equivalence between the labor of the organ and that of the machine—the equivalence of abstract, divisible labor time. And this equivalence meant that the performance of the machine and of the body could be subjected to similar strategies of regularization and even reproduction.

In his later work Marey looked at the ways in which the principle of energy conservation, governing the organ and its metabolism, could be combined with mechanics to explain the interrelationships between organs and the laws of movement of the whole organism. The work of the German mechanical engineer Franz Reuleaux provided him with the first systemic theorization of the machine, its laws of motion and composition. Reuleaux's science of the machine, which would later be associated with the discipline of mechanical engineering, was referred to as kinematics. In his 1875 work, *The Kinematics of Machinery*, Reuleaux outlined the minimal conditions for the creation of an effective mechanism. Kinematics, he stated, requires that motion be "of an absolutely defined nature," contingent on a predefined assemblage of parts and containable within certain parameters of space and time. To achieve this state of motion,

we need to "give beforehand to the parts which bear the latent forces, the bodies, that is . . . such arrangements, form and rigidity that they permit each moving part to have one motion only, the required one" (Reuleaux [1875] 1963,
35). Kinematics, in other words, needs stiff bodies.

Marey himself reaffirmed this principle when he stated that "strict relations
exist between the form of the organs and the character of the organs," between
the reversibility and predictability of function and the permanence of form (cited
in Rabinbach 1992, 92). What the rigidity of mechanical bodies made possible, as stressed by Reuleaux, was a certain restriction of movement. That is,
once bodies were presumed to be rigid, all that was left for the machine to
perform were geometric displacements, point-to-point translations, inversions,
and rotations. "We shall assume in the first instance that the bodies possess
complete rigidity and shall pay no attention to their size . . . so that only geometrical properties remain for us to consider" (Reuleaux [1875] 1963, 42).
Reuleaux thus established kinematics as a science of metric transformations.
At the same time, he precluded any consideration of nonmetric, continuous
transformations of bodies from the proper purview of mechanical science. In
the world of kinematics, form and function could be translated, exchanged,
substituted for in space and time—as long as form itself underwent no alteration. "Its province," he wrote, "is how to give the bodies constituting the
machine the capacity for resisting alteration of form." (ibid., 39).

Drawing inspiration from the science of kinematics, Marey's later work was
concerned with establishing the laws of bodily motion. Like Reuleaux, Marey
was interested in movements that could be easily translated into metric terms—
locomotion, the heart beat, electrical impulses, the acoustics of the ear, optics.
Physiological motion, he asserted, can be reduced to infinitely small instances
of time. In principle, it is infinitely divisible. Marey's theory of motion nevertheless presupposes an ideal limit to movement, a tangent of absolute stillness
where all curvature ultimately cancels out. The essence of motion can thus be
captured in the abstract instant where time reduces to space and movement
is frozen in a still frame. "All movement is the product of two factors: time and
space; to know the movement of a body is to know the series of positions which
it occupies in space during a series of successive instants" (cited in Rabinbach
1992, 94).

In this regard Marey was drawing self-consciously on the insights offered by the shift from photography to cinematography in the late nineteenth century. The relation between the two, he argued, was one of temporal decomposition and recomposition: whereas the photograph immobilized time into so many frozen instants, the cinematic image was able to reconstitute movement by bringing all these instants back together in rapid succession, blurring them together like so many flash cards. Implied here is the notion that time-instants are ultimately reversible, exchangeable, in and of themselves indifferent to change, just like points in space. In this way, for example, a film sequence of a horse in movement can be played backward without altering the form of the horse.

However, Marey was interested not only in the photographic and cinematographic image (dedicating numerous experiments to recording the movements of organisms), but also in what he called physiological time—the internal motions of the body, circulation, pulse, metabolism, and the electrical impulses of the muscles and nerves. Many of his inventions were designed to measure the internal rhythms of the body, just as the microscope had delivered up the visible inside of the body in the form of dissection slices. It is clear that Marey considered it possible, in principle at least, to recompose and thus regulate these internal time-instants, just as it had become feasible to reconstitute the moving, cinematographic image from the photographic still. In line with the visual arts, he suggested, it was time for the life sciences to move beyond the static arts of microsection toward a reconstructive surgery of mobile recomposition. Marey wrote: "We seem to have been traversing an immense gallery of mechanisms of greatly varied combinations . . . but everything here was mysterious in its immobility. The shift from organic structure to dynamics and the 'interplay of organs' was a shift to mobility and to 'motor function'" (cited in Rabinbach 1992, 91).

In this sense Marey envisaged the life sciences moving away from a paradigm of suspended animation (the organism viewed from the interior as a dissection slice or time still) toward a technique of cinematographic reanimation, where the suspended organ would again become mechanically functional. He was fascinated by the possibilities of mechanical reanimation and constructed numerous prosthetic organs and lifelike automata. And by the turn of the century such biologists as Alexis Carrel and Charles Lindbergh were experimenting with the idea that the actual biological organ could be suspended in time, transplanted and reanimated in much the same way as the photographic still.[2] Carrel and Lindbergh quite explicitly described their studies in organ culture as an exercise in suspending and recomposing the organ in time and space. Whereas dissection puts a stop to the livable life of the organ, and vivisection merely suspends it on the verge of death, the whole aim of organ transplantation is to suspend animation to then revive it elsewhere.[3]

The conceptual force of Marey's philosophy is evident in later twentieth-century approaches to biomedical technologies. A representative popular science text published shortly after the World War II transplant and prosthetics boom, for example, details the enormous technological advances that have occurred over the past two decades, while adopting a conceptual framework on bodily time and motion that barely differs from Marey's cinematographics of the organ (Longmore 1968). The machine and the organ obey the same principles of energy conversion, the author claims, the only difference residing in the infinitely greater complexity and efficiency of the biological machine. Moreover, it is this essential equivalence between the organ and the machine, between one process of energy conversion and another, function and form, that makes substitution possible. Organ transplantation, like prosthetics, is an art of "spare parts." One organ can substitute for another, just as one prosthetic equivalent can take the place of an organ, as long as the essential relations between form and function are preserved. As the text recounts the technical problems that have so far been encountered in the development of organ technologies, it becomes clear that any "alteration of form"—any morphological change in the organ to be transplanted—can only be registered as an obstacle or disturbance. As Reuleaux (1963, 35) had warned, "every motion . . . which varies from the one intended will be a disturbing motion." In other words the point of both prosthetics and whole organ transplantation is to realize a seamless translation of the organ in space and time—point-to-point substitution—while suppressing any other kind of change.

In order for this to occur, the organ itself must be suspended in time, its form solidified, its metabolism slowed down, at least until the act of transplantation is completed. In this sense advances in whole organ transplantation have been inseparable from mid-twentieth-century developments in the preservation of organs, either through whole organ perfusion (where the organ is constantly perfused with a solution) or hypothermic (above 0 degrees) cold storage—both of which aim to arrest the organ in time.[4] But it is not only the transplant organ that needs to be frozen. The science of organ transplantation also confronts the problems of tissue matching, immune reactions, and inflammation in the body of the recipient as so many disturbances to the smooth transplantation of rigid bodies in space, purely negative limitations that need to be overcome or simply suppressed (for example, through total body radiation or immunosuppressive drugs).

In short, all of these methods are intent on disarming the alterability of the organ, just as Reuleaux's treatise on kinematics can only begin once he has made the theoretical decision to abstract from the malleability of metals. As modes of biomedical intervention, prosthetics and organ transplantation need to arrest the process of morphogenesis to then work with the frozen morphological form. In this respect, as Canguilhem has pointed out again, the mechanistic vision of animation is entirely alien to that other branch of late-nineteenth-century biology, experimental embryology, which was precisely interested in understanding and intervening in the morphogenesis of form as process. Canguilhem (1992, 119) writes: "It was work in experimental embryology, above all, which led to the decline of mechanistic representations in the interpretation of living phenomena, by showing that the germ does not enclose a sort of 'specific machinery' ([Lucien] Cuénot) destined to produce this or that organ once it had been activated. . . . studies on the potentialities of the egg, following on from the work of [Hans] Driesch, [Sven Otto] Hörstadius, [Hans] Spemann and [Hilde] Mangold, made it clear that embryological development cannot be easily reduced to a mechanical model."[5]

ORGANOGENESIS: MODULATING MORPHOGENESIS

Following on from Canguilhem's work, what happens to this difference between the mechanistic and morphogenetic view of animation when experimental embryology lends itself to biomedical science, becoming (as in TE) an art of bodily reconstruction? What are the respective roles of morphological form and morphogenetic process in these different biomedical technologies? In one sense the difference is obvious: rather than suppress the body's responsiveness to change, TE aims to foster and work with the process of morphogenesis, in order to then modulate, reverse, or redirect it. TE is thus more concerned with the genesis of form—*organogenesis*—than the transplantation of already given forms. But how does it refigure the act of generation? And what specific modes of transformation does it deploy? Consider the relationship between form and morphogenesis in the following techniques currently under development.

In one such experiment cells are seeded and cultured in a collagen gel. The construct then grows and reorganizes the surrounding gel until it becomes superfluous. Gels have been used for growing soft-tissue constructs such as skin but are less successful at creating more solid forms because of the weak structural properties of the tissue. An alternative method, using three-dimensional porous biodegradable scaffolds, has been proposed as a way of overcoming the problem of structural weakness. Cells are seeded into the scaffold, gradually forming the extracellular matrix that provides the adhesions between cells. As the cells and extracellular matrix take on structural properties of their own, the scaffold slowly breaks down. One problem that arises with the use of such scaffolds is the low-level but protracted inflammation response it provokes in surrounding tissues. The problem of inflammation is sidestepped in a third method, which seeks to culture cells in such a way that they create and embed themselves in their own extracellular matrix without any outside support. The resulting sheets of cells are then folded, stacked, and rolled to obtain various morphological forms and densities of tissue. Of all the methods currently being developed, this one comes closest to the in vivo processes of organogenesis in that it does away with gels and scaffolds entirely.[6]

Once the process of seeding the cells has taken place, the tissue constructs are placed in a bioreactor, a machine that serves not only to culture cells in a sterile, growth-culture medium but also to subject them to various physical stimuli. The operative question can be posed as follows: In response to what forces and tensions, and at what threshold, will an ensemble of cells, defined by variable relations of adhesion or disconnection, fold into a particular morphological form and acquire particular cellular properties? Regenerative medicine works through the continuous variation of force fields, and it is from this level up that it attempts to determine the emergence of particular tissue qualities (density, compressibility, elasticity), properties, and forms (cell morphology and differentiation, organ morphology and structure). These forces might be biochemical, hydrodynamic, mechanical, or even electromagnetic in nature. Thus researchers have explored the different kinds of morphologies that emerge when a tissue construct is subject to static culture conditions: turbulent flow under rotation or laminar flow; various kinds of biochemical stimuli; the effects of waveforms of different shapes and frequencies; periodic compressive strain; and microgravity conditions (Freed and Vunjak-Novakovic 2000; Mejia and Vilendrer 2004). Other researchers have experimented with the properties of the scaffold (changing its shape, porosity, stiffness, and strength) to find out which are the most suitable for particular kinds of tissue growth (Sun, Darling, Starly, and Nam 2004; Sun, Starly, Darling, and Gomez 2004).

It is clear from these examples that the morphological form of the developed organ no longer plays the structural role that it does in earlier biomedical technologies such as organ transplantation and prosthetics. In these older technologies, form is presupposed or imposed from the outside—the technical act of implantation has nothing to do with the morphogenesis of the organ itself, which must remain unaltered in time, at least for as long as the act of transplantation is carried out. TE, however, is interested precisely in the continuity of morphogenesis in time. Far from requiring the solidification of the organ (its biological inertia), the technicity of the intervention works with and exploits the active responsiveness of living tissue, its power to affect and be affected and thus to change in time. Inasmuch as it incorporates a kind of "form," this is a gelatinous or porous protoform (the bloblike collagen gel and spongelike matrix), a form that is designed to be reabsorbed as the construct begins to take over.

The whole point, then, is not so much to *impose* form from the outside (the form is in any case designed to self-destruct), but rather to determine the threshold conditions under which an ensemble of cells, defined by certain relations, will *self-assemble* into a particular form and tissue, with particular qualities. The difference between these two methods can be likened to philosopher Gilbert Simondon's distinction between the mould (active form on formless matter) and the continuous modulation of force as modes of technicity. Simondon (1995, 45) writes: "To mould is to modulate definitively; to modulate is to mould in a manner that is continuous and perpetually variable."[7]

Much of the current work in TE remains at the experimental stage. But there is also a nascent theoretical literature that goes further than these ad hoc experiments and seeks to formalize the relationship between such parameter modulations and the morphogenesis of form. No doubt, the most sustained attempt to think through the implications of regenerative medicine can be found in the work of Donald E. Ingber, a biologist who combines a practical perspective on the science of biomedical reconstruction with a comprehensive theory of morphogenesis. Ingber characterizes his own work as an architectural or mechanical theory of morphogenesis, one in which the semiotic concepts of developmental biology (signals and transmission) are all refigured in terms of force relations. In Ingber's words (2003, 1397): "How does a physical force applied to the ECM [extracellular matrix] or cell distortion change chemical activities inside the cell and control tissue development? The answer lies in molecular biophysics; . . . it also requires that we take an architectural perspective and consider both multi-molecular and hierarchical interactions."

This is not any kind of architecture or mechanics of force, however, since Ingber is interested in continuous variations of tension rather than the divisible, extensive forces of classical mechanics. Working at the level of tissue morphology, he draws on the architectural theory of tensegrity to account for the morphogenesis of organic form. Developed by the architect Buckminster Fuller, the concept of tensegrity refers to structures that are maintained through the continuous transmission of tension rather than compression (as in the classic stone arch). It is a theory of structure that begins with continuous fields of forces rather than discontinuous forms, and capabilities (the potential for a spectrum of transformations) rather than intrinsic properties or qualities. In this way tensegrity architecture dispenses with the need for rigid bodies and predetermined materials.

According to its calculations, the structural stability of a form is not dependent on the rigidity of its constitutive parts but emerges as the geometric solution to a specific configuration of continuous forces and counterforces. Elastic bodies may thus yield rigid, stable structures, depending on the kind of force relations in play. Highly plastic, deformable cells (adult or embryonic stem cells) may be induced along specific pathways of development as a function of their adhesive relations. Crucial to tensegrity theory are the concepts of force fields, connective relations, "force transduction," and "action at a distance." In tensegrity structures there is no unit that is not composed of and traversed by connective interactions, no local event that is not responsive to global modulations of tension. Ingber thus understands morphological form as the emergent effect of connective interactions working across structural levels, within the cell, between cells and the extracellular matrix (adhesive interactions), and across the cell surface.

According to this model, changes of form, structure, and even cell fate are a function of field relations—the variable thresholds of tension traversing a field of connective relations and acting across levels or "at a distance"—and the continuous deformation of these relations give rise, at various thresholds of tension, to new forms and structures. In this respect Ingber's theory of cellular architecture is predictive, even if it eschews simple linear prediction. He is interested in determining, in mathematical terms, the parameter variations in which a given form can be modulated, the thresholds at which it will undergo a phase transition and mutate into another form. And it is from here that he derives his understanding of TE as a technical intervention.

All this suggests a very different perspective on the definition of regenerative medicine as an architectural or mechanistic practice of morphogenesis. For although TE certainly works with relations of force, it can no longer be described as mechanistic in the classical sense of the term. And while it marks a return to earlier traditions of biomedical engineering (as opposed to the semiotic models favored by the genetic revolution), it no longer involves the metric transformations of prosthetics and organ transplant technologies. TE explores continuous morphogenetic spaces and their capacity to engender a whole spectrum of bodies. It replaces the techniques of reproduction and substitution with an art of continuous modulation in which form is plunged back into process, becoming continuously remorphable. In a word, the bodily transformations induced by TE are *topological* rather than metric.

ON TRANSFORMATIONS: METRIC AND TOPOLOGICAL

The nineteenth-century mathematician Félix Klein established a method for distinguishing between different geometries as a function of the kinds of transformations they make possible. According to the mathematics of transformation, the pertinent question is no longer: What is the essence of a geometric figure? But rather: *What transformations is it capable of undergoing without changing in nature?* The key concept here is invariance under transformation: each type of space entails its own rules of invariance in relation to particular kinds of transformation and can thus be classified as a function of these rules. The interest of the mathematics of transformation lies in its attention to the *capabilities* of space; that is, a geometrical figure is defined by its responsiveness to certain events—the transformations it is capable of performing or undergoing—rather than a set of static properties (DeLanda 2002, 18).

In Euclidean space, for example, which could perhaps be characterized as the least capable of spaces, figures can undergo transformation without changing in nature, as long as the metric properties of distance, length, and degree are preserved. Metric space is defined by the group of so-called rigid transformations—rotations, inversions, and translocations—that move bodies without altering their morphological shape or responses to force-relations. But nineteenth-century mathematicians were prolific in inventing geometries that escaped the metric invariants of Euclidean space. In affine geometry, for example, the circle and the ellipse become continuous, and in projective geometry all conic sections merge into the same field of variation. However, by far the least rigid, most pliable of geometries is afforded by topological space.

In topological space metric invariants are inoperative, as are the invariants of all other geometries such as projection and conic sections. In absolute topological space the inside is continuous with the outside, the left is reversible into the right, and morphologies continuously morph into each other. If we can continue to speak of the "point" in topological space, we would need to say that it is in continuous transformation, each point morphing into the other at infinite speed (alternatively we could say that the point does not exist as such, that the continuity of transformation precedes the immobility of the point). In topological space the rigid, discontinuous bodies of metric space become indiscernible, indefinitely transformable into each other. What topological transformations leave unchanged are simply connective relations—viscosity as an abstract marker of togetherness or adhesion.

These relations might connect points or forces, but in either case the connections they establish operate independently of metric notions of extensive distance or degree (mathematicians use the concept of proximity without distance to define these relations of nonmetric togetherness or cohesion). Imagine contorting, stretching, and pummeling a doughnut into any shape possible as long as it conserves its morphological cohesion (and points of noncohesion, such as the hole). Imagine a viscous blob being subjected to the most extreme variations of force, compressed and stretched in every direction, without losing its generic blobiness. In topological space the whole spectrum of possible shapes, qualities, and consistencies are all equivalent under transformations that preserve the topological property of connectedness. What counts here is not the rigid form of the doughnut but doughnut morphogenesis as a field of continuous variation, operating within certain parameters of connectivity. Topological space recognizes difference only when it involves a change in the nature of connective relations. At this point we move from one topological *neigh-borhood* to another, from one force field of variation to another.

But what of the relationships between these different spaces or geometries? Klein's work is of interest not only because he sought to categorize different geometries as a function of invariance but also because he established an onto-genetic or generative relation between them. According to Klein's formulation of the problem, the least continuous of spaces (metric space) can be engendered from the most continuous of spaces (the topological) through the gradual emergence of discontinuities and corresponding loss of transformability (in modern mathematical terms this is described as a loss of symmetry or symmetry-breaking). Klein's mathematics of transformation thus provides the first *onto-genetic* theory of space: "Metaphorically, the hierarchy 'topological-differential-projective-affine-Euclidean' may be seen as representing an abstract scenario for the birth of real space. As if the metric space which we inhabit and that physicists study and measure was born from a non-metric, topological continuum as the latter differentiated and acquired structure through a series of symmetry-breaking transitions" (DeLanda 2002, 26).

All this suggests a certain conceptual affinity with embryology, that other nineteenth-century theory of generation. It is not surprising then that embryology, the scientific discipline that is most concerned with the process through which biological form comes into being, should have invented the concept of the *morphogenetic field*. Developed in the late nineteenth and early twentieth century, the morphogenetic field thesis allowed embryologists to account for the nonmetric relations of difference and resonance (action at a distance) that seemed to animate the early moments of embryogenesis. They noted, for example, that although no precise spatialization of the future organism could be found in the fertilized egg, the morphogenetic field was emphatically not devoid of differences. These differences, however, were of an intensive or non-metric nature—continuous variations of field intensity (gradients), fuzzy neighborhoods defined by field resonances or actions at a distance (polarities), and sheets of migrating cells moving at different speeds or dividing at different rates. It was from the encounter between these neighborhoods of nonmetric difference, they argued, that the first spatializing movements of the early embryo came into being, producing the foldings and invaginations that would only gradually give rise to a recognizable organism. In their conceptualization of the morphogenetic field most embryologists appealed to the topological language of force fields and resonance, but few attempted to formalize these nonmetric relations in mathematical terms.

One of the first theoretical biologists to explicitly think through the implications of transformation mathematics for the life sciences was D'Arcy Thompson. In his 1917 study on the morphogenesis of organic form, *On Growth and Form*, Thompson begins by identifying the mathematical theories that inform the different disciplines within the life sciences. While genetics, with its interest in discontinuous mutation events, implicitly relies on the mathematics of substitution groups, he suggests that the developmental and morphogenetic notion of variation has more in common with the mathematics of transformation. In the last chapter of his book, "On the Theory of Transformations, or the Comparison of Related Forms," Thompson develops a comprehensive theory of morphogenetic transformations as a way of accounting for the continuous variability in shape and structure that can be found in organic life. The interest of such a method, he argues, lies in its capacity to reproduce the various effects of force or strain in the morphogenesis of form, thus replacing a typology of essences with a grid of continuous morphological transformations. "In a very large part of morphology, our essential task lies in the precise comparison of related forms rather than in the precise definition of each; and the deformation of a complicated figure [becomes means of comparison]" (D. Thompson [1917] 1992, 271).

Having said this, the parameters that Thompson chooses to explore are restricted to the Cartesian x, y coordinates and their possible curvatures. These morphological transformations are performed as follows: after outlining the form of an organism in a system of x, y coordinates, the system is subject to a uniform strain, thereby producing a corresponding deformation in the form of the organism. As Thompson (ibid., 272) points out himself, this method is akin to the projective transformations used in cartography, where a curved surface can be transferred, through deformation, onto a flat surface and vice versa. The choice of such a method implies a corresponding restriction in the kinds of deformations it is able to perform. Thompson's transformations effectively presuppose a certain cartography of organic form, the existence of empirical, taxonomic discontinuities between morphological kinds that therefore limit the possibilities of further transformation: "We shall strictly limit ourselves to cases where the transformation necessary to effect a comparison shall be of a simple kind, and where the transformed, as well as the original coordinates shall constitute a harmonious and more or less symmetrical system. We should fall into deserved and inevitable confusion if, whether by the mathematical or any other method, we attempted to compare organisms separated far apart in Nature and in zoological classification" (ibid., 273).

The shortcomings of D'Arcy Thompson's method are later explored by the theoretical biologist J. H. Woodger (1945), who notes that his transformations are restricted to shape and thus presuppose a certain rigidity of space. Thompson's method allows him to trace the continuous variation of one related form into another through the curvature of space but stops short at the visible, taxonomic differences between kinds. For Woodger the limitations of Thompson's method lie ultimately in his near exclusive focus on the already developed morphology of the adult organism. Having restricted himself to the already completed forms of the adult body, Thompson is unable to account for the morphogenesis of form itself. A more comprehensive theory of morphology, argues Woodger, would need to become genetic (by which he means generative), and in order to do this, it would be necessary to begin with the first moments of embryogenesis, where discontinuities of form and shape are not so self-evident. Although Woodger does not explicitly draw this conclusion, his critique of Thompson's work suggests that a truly ontogenetic theory of morphogenesis must move from the relatively restricted though nonmetric space of projective transformations to the continuous space of topology.

This is precisely the project that will later be taken up by the mathematician René Thom in his celebrated 1975 study of morphogenetic models. In Thom's own words the aim of this study is to give mathematical sense to the embryological concept of the morphogenetic field, using the theory of topological transformations. *"The fundamental problem of biology,"* he claims, *"is a topological one, for topology is precisely the mathematical discipline dealing with the passage from the global to the local"* (Thom 1975, 151). Biochemical and genetic models, he argues, have failed to address the problem of morphological stability and form. What Thom proposes instead is a comprehensive geometric theory of morphogenesis, one that does not so much dispense with the insights of biochemistry as incorporate them within a larger dynamic perspective. In this way both local and collective events become interpretable within the one field of dynamic relations: "In place of explaining the overall morphogenesis by modifications of the cellular ultrastructure, we try to explain the cellular ultrastructure by dynamical schemes similar to those of global morphogenesis, but on the cellular level" (ibid., 156).

Beginning with the morphogenetic field as a space of absolute continuity, what René Thom is able to theorize is not only the transformability of certain classes of adult morphologies (à la D'Arcy Thompson), but more important the embryogenesis of the morphological itself. In this way he attempts to delineate a general theory of organic morphogenesis, where the generation of bodies is envisaged, following Félix Klein's method, as a successive unfolding of geometries or spaces. This unfolding moves from the most transformable, plastic, continuous of spaces (the topological or the embryological) to the relatively rigid forms of the adult body. Or rather, in the language of modern nonlinear mathematics that Thom also uses, the developing organism emerges from the topological (the most symmetric of spaces) through successive bifurcation or symmetry-breaking events into spaces that are less transformable but more complex in their actualized forms.

Thom thus envisages the emergence of the first spatial structures of the body (inside and outside surface, left and right, and so on) as thresholds at which the morphogenetic field is compelled to actualize its continuous intensities in the form of abrupt, discontinuous actualizations of difference (the alternative is chaos!). He describes these events as a series of "catastrophes" (in the sense that they elude simple or linear prediction) and delineates seven generic spatializing events in the morphogenesis of the organism: the fold, the fringe, the dovetail, the butterfly, the umbilic hyperbolic, the elliptic umbilic, the parabolic umbilic (*le pli, la fronce, la queue d'aronde, le papillon, l'ombilic hyperbolique, elliptique, parabolique*). Thom's program has more recently been taken up by so-called structural biologists Brian Goodwin and Gerry Webster (1996), although this time with the important contribution of topological computer simulations.

If we now try to conceptualize the technical process at work in TE, it could be said that it aims to reproduce the topological space of the early embryo in order to then regenerate the successive geometries of the body otherwise. What Thom proposes as a topological theory of embryogenesis here becomes a reproducible technical procedure: the three-dimensional culturing of a field of relatively uncommitted cells in various states of tension and connectedness. In the first instance, then, TE plunges the body back into the morphogenetic field, where inside and outside morph into each other, where the salient differences are not yet organs and parts but organ fields and continuous variations of force. This return is most extreme when the cells used are the highly plastic, mutually transformable stem cells of the early embryo.

In this case TE collapses the body back into its own field of embryolog-ical transformations in order to provoke it along different pathways of devel-opment. Or rather, it makes the body indefinitely regenerable, available to its entire field of possible deformations, monstrous or otherwise. As the author of one recent overview put it: "The fundamental principle of reparative med-icine which governs our efforts to regenerate differentiated tissue is to organ-ize the reparative circumstances to recapitulate selected aspects of embryonic developmental sequence, including attempts to mimic the embryonic microen-vironment in which tissue initiation, formation, and expansion take place" (Caplan 2002, 307). TE thus creates a kind of topological hologram of the body, a virtual space in which the entire field of bodily transformations can be mate-rialized in vitro, if not lived in vivo! As a morphological intervention, it is akin to recent developments in topological architecture and computer-assisted design, which allow the designer to regenerate (both virtually and actually) a whole spectrum of forms and shapes from the one topological neighborhood (Cache 1995; Di Christina 2001).

In principle, this return can be repeated ad infinitum, so that TE is capable of perpetuating embryogenesis, of reliving the emergence of the body over and over again, independently of all progression. Here it is not only spaces, forms, and bodies that become continuously transformable, but also the divisible instants of a chronological lifetime, so that any one body can be returned to or catapulted into any point in its past or future, and into any past or future it *could have* and *could still* materialize. In other words TE not only seeks to "return" the body to nonmetric *space* but also to nonmetric *time*—and to reca-pitulate the various chronologies of morphogenesis from here.[8] In principle, then, the adult body will be able to relive its embryogenesis again and again—including those it has never experienced before.

MODES OF ORGAN ASSEMBLY:
FROM STANDARDIZED TO FLEXIBLE PRODUCTION

The rise of regenerative medicine, of course, is not simply attributable to the development of novel technical possibilities, but rather participates in a more general shift in modes of production. This is the shift that takes us from Fordist modes of mass, standardized reproduction into a post-Fordist economy of flexi-ble, nonstandardized production; from the reproduction of the standardized form

(the norm) to the regeneration of the transformable, the emergent, or the anom-
alous. In effect, the post–World War II era inaugurated the mass industrialization of prosthetics as well as such biomedical procedures as blood transfusion,
tissue transfer surgery, and later organ transplantation. In his 1994 study of
the Taylorist/Fordist revolution in industrial production, the sociologist Benjamin Coriat offers a detailed overview of the time and motion imperatives of
massified labor. In the first instance, he notes, Taylor's reforms sought to reduce
complex chains of production to a series of elementary time and motion units,
each of which could be easily reproduced by an unskilled worker or a simple
machine. With the introduction of the assembly line, Fordism extended this
process of rationalization to the industrial object itself. The standardized Fordist
object was one that satisfied strict norms of quality control, equivalence, and
reproducibility, thus facilitating its mass processing and distribution. It was
designed to preserve a certain self-identity, to maintain value for a guaranteed
amount of time without loss or fatigue (Coriat 1994, 45–84).

In the arena of biomedicine, I suggest, such methods of standardized collection, processing, and distribution required a number of specific innovations:
the most essential of these was the creation of the first centralized blood and
tissue banks during World War II, which in turn was inseparable from developments in the storage of tissues (cryopreservation), knowledge in immunology, and tissue typing. Later on, when organ transplant became a more common
procedure, strict regulatory guidelines would be issued in order to maximize
organ preservation, reduce circulation and transport times, and standardize both
organ storage and packaging.[9] In a general sense the public tissue bank was set
up as a repository in which units of exchangeable organ-time could be stored,
classified, matched, and identified, to ensure the greatest possible interchange-
ability of parts, at the lowest possible cost in terms of organ waste, loss, or fatigue.
Such methods of standardization imply a certain indifference to the actual materiality of the functional organ replacement—prosthetic or biological—or rather
they attempt to create the conditions under which the difference between the
organic and the mechanical becomes effectively negligible.

Perhaps not incidentally, then, the sought-after characteristics of both the
transplantable organ and the prosthetic body part are encapsulated in the post–
World War II definition of the *medical device* (a term that was adopted in both
North American and European regulations). The U.S. Food and Drug Administration, for example, defines the medical device as "an instrument, appara-

tus, implement, machine, contrivance, implant, in vitro reagent, or other similar or related article . . . which does not achieve any of its primary intended
purposes through chemical action within or on the body of man or other animals and which is not dependent upon being metabolized for the achievement
of any of its primary intended purposes," a definition that clearly excludes any
implant (synthetic or biological) that might interact with the body in any way
after insertion.[10]

The philosopher Gilles Deleuze (1993, 19) has offered the following suc-
cinct overview of the key differences between standardized and emergent production: "The goal is no longer defined by an essential form, but reaches a pure
functionality, as if declining a family of curves, framed by parameters, inseparable from a series of possible declensions or from a surface of variable curvature that it is itself describing. . . . As Bernard Cache has demonstrated, this
is a very modern conception of the object: it refers neither to the beginnings
of the industrial era nor to the idea of the standard that still upheld a sem-
blance of essence and imposed a law of constancy ('the object produced by
and for the masses'), but to our current state of things, where fluctuation of
the norm replaces the permanence of a law."

Although still in its experimental stages, then, post-Fordist biomedicine can
be described as the paradigm in which the reproduction of standardized form
gives way to the continuous regeneration of the organ's field of transformation. Compare, for example, the different modes of storage, fabrication, and
assembly associated with organ transplantation (and prosthetics) on the one
hand and TE on the other. Essential to the large-scale implementation of organ
transplants are precise techniques and protocols for cooling, freezing, packaging, and transportation, all of which are designed to maintain the form and
function of the organ in time. Of course, regenerative medicine does not dispense with these methods; it is in fact intimately dependent on the ready avail-
ability of such standardized tissue and organ supplies (Naughton 2002). But it
also introduces an important new element into the process of fabrication itself—
the bioreactor. While the bioreactor is in one sense an incubator that can be
used to store and transport a tissue construct, its purpose is also to provide the
conditions under which a tissue can be modulated, deformed, continuously
remolded. As an instrument of fabrication, what the bioreactor promises to
deliver is not a standardized equivalent, but a whole spectrum of variable tissue forms, all of which may be generated from the one tissue source.

Such novel modes of production have called for their own methods of abstraction and simulation. It is in the field of architecture that computer-assisted topological modeling has been carried the furthest. Architects have now moved on from the virtual simulation of topological spaces to their actual fabrication. The design theorist Bernard Cache thus distinguishes between two stages in the application of computer-assisted modeling to architecture. The first generation systems of computer design merely scaled-up traditional hand-designed architecture by using the standard geometric combinations of translation and rotation along with approximated curves (Beziers or Splines). The second generation of design, however, introduced parametric functions, fractal geometry, and continuous topological modeling, in ways that had not previously been accessible. Cache (1995, 88) writes: "Now, we can envisage second-generation systems in which objects are no longer designed but calculated. The use of parametric functions opens two great possibilities for us. First, this mode of conception allows complex forms to be designed that would be difficult to represent by traditional drawing methods. Instead of compositions of primitive or simple contours, we will have surfaces with variable curves and some volumes. Second, these second-generation systems lay the foundation for a nonstandard mode of production. In fact, the modification of calculation parameters allows the manufacture of a different shape for each object in the same series. Thus unique objects are produced industrially."

In comparison, computer-aided modeling for use in TE is only just beginning to get off the ground.[11] But already biologists and computer scientists are looking at the ways in which topological models might be adapted to this purpose.[12] In an important sense these theorists are not imposing an "abstract" model onto a resistant material; rather, they are merely acting on the results of experiments carried out in the actual modulation of tissue constructs, experiments that have highlighted the extreme mutability, unexpected recalcitrances, and peculiar generativity of the living biological.

Given these essential differences, it is not surprising that one of the major problems confronting the emerging field of regenerative medicine is that of finding some kind of fit between its extremely lively constructions and the standard specifications set down by federal regulatory agencies. So far, most companies working in TE have attempted to commercialize their tissue constructs by defining them as traditional "medical devices," a definition that implies some degree of stability, reproducibility, and standardized form. But as one commentator on the regenerative medicine industry has pointed out, the TE construct only works if it continues to grow and respond to surrounding tissue after implantation (Naughton 2002). Its productivity, in other words, is dependent on its continued ability to self-transform, to grow, to morph, in ways that are not easily predicted. It is here, perhaps, that lies the essential novelty of TE as a form of technicity, but also its potential for breakdown, accident, or error. One of the dangers that stem cell scientists in general have still to properly confront is that of excess tissue mutability. There is the distinct possibility, in particular, that the extremely plastic, mutable cells of the early embryo may end up proliferating too well, giving rise to cancerous growths rather than restoring health.

Here, it could be argued, lies the particular kind of structural breakdown that characterizes the post-Fordist, as opposed to Fordist, techniques of biomedical production. Where the machine body of the industrial era was plagued by the problems of fatigue, depletion, or entropy (Rabinbach 1992), the postindustrial body is more likely to be overcome by a *surplus productivity that is indistinguishable from a surplus of life*—that is, crises of overproduction or the dangerous, excessive vitality of cancer.

MODES OF ANIMATION: RETHINKING BODILY TIME

The medical anthropologist Margaret Lock (2001, 291–94) has compared the biomedical understanding of animation with Christian medieval imaginaries of bodily fragmentation and resurrection, describing modern-day practices of organ transplantation as a kind of mutualized, democratized instantiation of such medieval beliefs. These beliefs are not as far removed from the life sciences as they might at first appear. Early pioneers of cryobiology envisaged the whole process of freezing, banking, and thawing tissues and organs as a quite literal form of suspended animation, followed by resurrection (Billingham 1976). Many of the first experiments in the field were conducted in parallel with studies on hibernation, freeze-resistance, and the capacity for suspended animation in certain kinds of animal species. And these studies are themselves indebted to late-eighteenth-century debates on death and resurrection, where resurrection was envisaged as a problem of relevance to both theology and the natural sciences (Rensberger 1996, 248–55).

But whatever the interest of such comparisons, the modus operandi of organ transplants would be more profitably compared with the processes of transubstantiation, suspense, and resurrection that Marx saw at work in the transformation of human labor time (organ-time) into the abstract, exchangeable labor time of the fetishized commodity form:

> Living labor must seize on . . . things, awaken them from the dead, change them from merely possible into real and effective use-values. Bathed in the fire of labor, appropriated as part of its organism, and infused with vital energy for the performance of the functions appropriate to their concept and to their vocation in the process, they are indeed consumed, but to some purpose, as elements in the formation of new use-values, new products. . . . (Marx [1867] 1990, 289–90)

> . . . insofar as labor is productive activity directed to a particular purpose, insofar as it is spinning, weaving or forging, etc., it raises the means of production from the dead merely by entering into contact with them, infuses them with life so that they become factors of the labor process, and combines with them to form new products; . . . (Ibid., 308)

> While productive labor is changing the means of production into constituent elements of a new product, their value undergoes a metempsychosis. It deserts the consumed body to occupy the newly created one. But this transmigration takes place, as it were, behind the back of the actual labor in progress. (Ibid., 314)

Organ transplantation, in other words, might be compared with the process by which the time-motion capacities of the laboring organ are abstracted from the worker's body and transformed into interchangeable units of time and money. Except here what is at stake is a quite literal abstraction of the biological organ itself. What is common to both forms of labor is the movement by which the livable time of the organ (time that can be measured as an expenditure of force, or negatively as fatigue) is extracted from the unicity of the body, stored, frozen, transformed into an exchangeable equivalent, circulated across generations and bodily boundaries and (in the case of xenotransplantation experiments) even species. In the course of this process an expenditure of mechanical force becomes strictly equivalent to the labor of the body. What mass commodity production and the mid-twentieth-century sciences of the organ share is a common understanding of time: one in which all movement can be reduced, in the last instance, to a series of interchangeable, equivalent presents—abstract organ-time.

This is where regenerative medicine is strikingly different. If organ transplant medicine needs to maintain life in a state of suspended animation, regenerative medicine, it might be argued, is more interested in capturing life in a state of perpetual *self-transformation*. Life, as mobilized by regenerative medicine, is always in surplus of itself. This is not to argue, of course, that regenerative medicine dispenses with the methods of tissue and organ preservation—such methods are essential once a tissue construct has been created—but what it *works* with is the body's capacity to elude such moments of suspended animation and frozen form. If it "reproduces" anything, it is not this or that generic form but the process of transformation itself. What regenerative medicine wants to elicit is the generative moment from which all possible forms can be regenerated—the moment of emergence, considered independently of its actualizations.

In what sense, then, are we to understand the term "moment"? In contrast to the frozen present of exchangeable organ time, what regenerative medicine requires is a strictly nonmetric, nondivisible understanding of time. Or rather, it suggests that the "instant," even when reduced to an extreme point of suspense, is always undercut by the continuity of transformation, change, or becoming. The instant, in other words, is never contained in itself, never present to itself, but (following Deleuze) perpetually about to be and already past, about to emerge and already subsided, *about to be born and already born again*.[13] To instantiate this generative moment in technical terms, regenerative medicine needs to cultivate the process of embryogenesis in such a way that it maintains its full spectrum of transformative possibilities. It requires a state of embryonic being that never grows up into this or that particular organism: a process of *self-perpetuating, unactualized, and unfinishable embryogenesis*. This is quite literally what biologists are attempting to accomplish when they culture and "immortalize" an embryonic stem cell line.

Thus far, I have explored the implications of tissue engineering and regenerative medicine for understandings of the spatial and temporal transformability of bodies. In the next chapter I widen my analysis to the problematic of bodily generation, reproduction, and regeneration. This problematic requires us to look in greater detail at the development of stem cell science—and the challenge it poses to many of modern biology's founding assumptions about the reproduction and regeneration of bodies. Here I am interested in comparing the different understandings of cellular and bodily generativity that are implied in the techniques of reproductive and regenerative medicine. I also explore the specific forms of economic and social productivity to which these technologies have been subject. If certain aspects of the reproductive sciences have long been included in commercial tissue markets, what forms of capitalist investment characterize the emerging field of stem cell science? What is the relationship between the embryoid bodies produced by stem cell science and the highly financialized modes of capital accumulation that proliferate in contemporary market economies? What, finally, will be the place of reproductive and regenerative labor in the emerging tissue economies of the twenty-first century?

□ 5 □

LABORS OF REGENERATION

Stem Cells and the Embryoid Bodies of Capital

> Finance, the management of money's ebbs and flows, is not simply in the service of accessible wealth, but presents itself as a merger of business and life cycles, as a means for the acquisition of self. The financialization of daily life is a proposal for how to get ahead, but also a medium for the expansive movements of body and soul.
>
> Randy Martin, *Financialization of Daily Life*

THE FIELD OF STEM CELL RESEARCH HAS A COMPLEX HISTORY THAT CAN BE traced back to multiple sites of origin. One of these is certainly the field of human reproductive medicine. In a 2001 article on the subject, R. G. Edwards, one of the biologists responsible for the first successful experiments in in vitro fertilization, argues that the first attempts to culture embryonic stem (ES) cells took place in his laboratories in the early 1960s, long before the idea of using ES cells as a transplant tissue had emerged. Edwards recounts how, in the course of routine lab studies, his students attempted to culture outgrowths of cells derived from very early-stage embryos, prior to their implantation in utero. The properties of these cell lines were unexpected. First, they displayed "immense properties of long-term survival" (Edwards 2001, 349). Like cancerous cells, it appeared that they were able to reproduce themselves indefinitely, without aging. Moreover, the cell lines were uniquely plastic and transformable, producing a whole spectrum of differentiated cell types, including blood, muscle, connective tissue, and neurons. It would be three decades before these same cell lines would be again cultured and formally characterized as ES cells, but by this time the cells had moved out of the realms of the experimental curio and established their place at the center of a whole new scientific and clinical enterprise—that of stem cell science (itself incorporated into the larger discipline of "regenerative medicine").

Today, reproductive medicine and stem cell science communicate with each other across a series of complex scientific, technical, and institutional interfaces (Franklin 2006). Many of the key practitioners of IVF are now involved in stem cell science; and the fortunes of IVF and stem cell science are also united by an intense and necessary traffic in tissues, since most of the current embryonic stem cell lines have been established using cells from IVF-derived embryos. As stem cell therapies begin to enter clinical trial, several new research centers in the United Kingdom and elsewhere have been specifically designed to bring IVF clinics and ES line derivation facilities together (ibid.).

Yet reproductive medicine and stem cell science differ in fundamental respects. This difference is first of all of a technical nature and concerns the actual generativity of cells. Beginning with the same materials, reproductive medicine and stem cell science seek to elicit different responses from the experimental life forms they culture in vitro, and ultimately offer different understandings of the nature of bodily generation itself. Even as it seeks to standardize and scale up the time units of reproduction, reproductive medicine remains within the parameters of sexual, germinal transmission—the paradigm established by the German biologist August Weismann in the late years of the nineteenth century. Stem cell science is still in its initial phases. Yet it already offers an entirely different perspective on the possibilities of cellular regeneration—one, I suggest, that threatens in the long run to displace the Weismannian paradigm as a reference point for thinking about biological generativity.

A key difference also lies in the kinds of institutional, political, and economic contexts in which reproductive and regenerative medicine participate. These contexts fundamentally inflect the value with which an experimental life form is endowed, confining its generative promise within certain limits and forcibly liberating it from others. Reproductive and regenerative medicine, I argue, participate in different kinds of "tissue economies," or rather function as different phases or moments of the one, highly stratified economy (Waldby and Mitchell 2006). It is the differences between these two moments that I explore throughout this chapter, to then understand the ways in which they work together. In particular, I am interested in the unique context in which IVF and stem cell science have developed in the US—where a highly deregulated market in privately funded scientific research and services exists side by side with an often intensely prohibitive stance on the part of the federal government (Green 2001; Jasanoff 2005).

In her detailed study of reproductive medical services in North America, cultural anthropologist Charis Thompson provides a useful set of premises for thinking through the politics of what she calls the "biomedical mode of reproduction." Her book begins with the observation that "U.S. biomedicine makes both production and reproduction part of the economy" but also points to the vast array of legislative and ethical limits that seek to curb the commodification of particular human tissues, particularly the embryo (C. Thompson 2005, 250). In the first part of this chapter I attempt to elaborate on Thompson's work, and perhaps complicate her conclusions, by looking at the areas of reproductive science in which the complete standardization of reproductive labor and the commodification of tissues are *not* prohibited. The apparent sacralization of the embryo evident in human reproductive medicine, I argue, needs to be contextualized within the history of North American agricultural reproductive science, where the mass production of reproduction has been fully implemented over the past few decades.

Far from being restricted to the agricultural industry, however, such practices are rife—if less formalized—in the arena of human reproductive medicine, which frequently finds itself resorting to the reproductive labor of various female underclasses to meet demand for scarce tissues (Dickenson 2001; Waldby and Cooper 2007). I also argue that the emerging field of stem cell science is being integrated into an entirely different process of accumulation, one that is irreducible to either (organic, human) production or reproduction in the Marxian sense.[1] My purpose in the second part of this chapter is to explore the specific modes of economic valuation that characterize the field of stem cell research in the United States. What is taking shape here, I argue, is something like an embryonic futures market, incorporating but not dispensing with the massified production of reproduction.

REPRODUCTIVE MEDICINE: HUMANIZING AGRICULTURE ·

The history of reproductive medicine dates back to the beginning of the twentieth century, when reproduction was identified as a subfield of the agricultural sciences. In her illuminating historical study of the area, the science historian Adele Clarke (1998) has recounted how mammalian reproduction emerged as a distinct area of research at about the same time that genetics and embryology split off into separate disciplines. This split, argues Clarke, occurred at a time when agriculture was becoming increasingly professionalized, mechanized, and organized along industrial lines. The principles of industrial production would henceforth be applied to everything from the seed (Kloppenburg 1988) to animal reproduction, giving rise in the process to a sizable new scientific literature dedicated to the rationalization of agriculture. A fundamental text in the creation of reproductive science, argues Clarke (1998, 68), can be found in F. H. A. Marshall's *Physiology of Reproduction*, published in 1910.

This text not only managed to establish mammalian reproduction as a field of study distinct from development and the transmission of heritable traits, but it also identified the processes of specific interest to the reproductive sciences. These included fertility, fertilization, estrous cycles and pregnancy diagnosis—processes that would soon be translated into such practical applications as artificial insemination, sex preselection, and twinning (ibid., 45). It was thus in the context of industrialized livestock production that reproductive medicine in general, and indeed gynecology, first emerged as distinct areas of scientific research. In the words of its founding father: "Generative physiology forms the basis of gynaecological science and must ever bear a close relation to the study of animal breeding" (Marshall 1910, 1, cited in Clarke 1998, 71).

The North American cattle industry of the 1970s was the first to deploy reproductive technologies on an industrial scale and was thus largely responsible for standardizing its procedures. The first of these new procedures involved the use of hormone injections to synchronize estrous cycles so that the donor and the recipient cow ovulate at the same time. Another development was superovulation: the use of hormone therapy to stimulate the production of an abnormally large number of eggs, which are then fertilized, usually by artificial insemination, and flushed from the body of the donor animal using a catheter. Finally, a crucial step in the development of commercial-scale embryo transfer, as for organ transplants, was the development of cryopreservation techniques making it possible to store, transport, and thaw embryos between the moment of donation and the moment of transfer.

As in any other area of Fordist manufacture, the aim of these procedures is to increase the production of relative surplus value (in milk and meat) by getting the most out of each unit time of reproductive labor. In line with the standard rules of assembly line production, animal reproductive science seeks to eliminate unproductive (or rather *unreproductive*) time, by extending the fertility of animals beyond their naturally fertile years or freezing and preserving embryos for later use; to maximize the production of surplus value, by augmenting the number of eggs produced at one time (superovulation) and twinning; to do away with obstacles of space and time, by freezing and transporting gametes and embryos; to overcome barriers to the circulation of reproductive materials through the use of artificial insemination, tissue transfer methods, and in vitro fertilization; and finally to standardize reproduction and eliminate mistakes using dissection microscopy and preimplantation genetic diagnosis. The overall effect of these technical interventions is to collapse the chronicity of genealogical time, rendering each moment of the reproductive process exchangeable against any other, and to transgress the boundaries between bodies and lineages. Hence, as several commentators have noted, some of the most disquieting effects of the new reproductive sciences—in particular, the suppression, confusion, or reversal of generational time—were evident in the field of animal biology long before their human applications brought them to the attention of bioethicists (Correa 1985, 74–76; Landecker 2005).

Reproductive medicine in the livestock industry, then, is a perfect instance of what Clarke has described as the application of Fordist methods of mass production to animal breeding:

> The reproductive sciences between 1910 and 1963 constituted a modernist enterprise par excellence. Modern approaches sought universal laws of reproduction toward achieving and/or enhancing *control over* reproduction. During the modern era, the reproductive processes focused on most intently by reproductive scientists and clinicians included menstruation, contraception, abortion, birth, and menopause; agricultural reproductive scientists also focused on artificial insemination. Control over reproduction was and still is accomplished by means of Fordist mass production-oriented emphases on the rationalization of reproductive processes, including the production and (re)distribution of new goods, technologies, and health care services that facilitate such control. (Clarke 1998, 9–10)

Indeed, the kinds of methods initiated by the U.S. cattle industry during the 1970s might be seen to represent the culminating step in Fordist mass production, since they not only apply the imperatives of mass commodity production to biological reproduction, but they also intervene in the temporality of reproduction—suspending, speeding up, and rearranging units of time—to a degree that had not hitherto been possible.[2] In this sense the U.S. meat industry seems to have been one sector in which the decline in Fordist productivity was offset by the recourse to acceleration.[3] In the process the mass mechanization of agriculture gave rise to a whole spin-off market in alienable, commercially evaluated reproductive tissues, which in turn engendered a demand for businesses specializing in the sourcing, banking, and exchange of tissue, along with the actual service of administering artificial reproduction.

What then is the relationship between agricultural reproductive medicine and the whole spectrum of human reproductive technologies that came onto the scene in the 1970s? Are we essentially dealing with the same phenomenon, as suggested by Marshall's original identification of gynecology with animal reproductive science (Clarke 1998, 71)? And are assisted reproductive technologies (ARTs) a form of industrial meat production applied to women, as such feminist critics as Gena Correa (1985) have forcefully argued?

Certainly the technological kinship between agricultural science and human reproductive medicine is undeniable. The field of human reproductive medicine, in its current form, emerged in the late 1960s and early 1970s, as a result of a number of key developments in surgical methods and hormone treatment (Wood and Trounson 1999). However, most of the fundamental techniques involved in the IVF process were invented in the context of animal reproductive science and date as far back as the late eighteenth century (Biggers 1984). The first (unsuccessful) experiment in the in vitro fertilization of animal ova was made in 1878, at a time when the fundamental processes involved in fertilization were just being discovered. In 1890 one of the pioneers of reproductive biology, Walter Heape, demonstrated the possibility of embryo transfer by flushing fertilized rabbit eggs from the fallopian tube and transferring them to a surrogate mother. There followed a number of attempts to replicate these earlier experiments. In 1934 the biologist Gregory Pincus announced that he had successfully performed in vitro fertilization of rabbit ova, although the results were later thought to be equivocal. It was only in the late 1950s that the feasibility of in vitro fertilization in rabbits would be incontrovertibly demonstrated, following a series of minute but key improvements in embryo transfer and culture, the collection of ova, and the understanding of reproductive cycles. It was this wave of developments, along with advances in the freezing of embryos and eggs, that would transform human reproductive medicine into a viable procedure.

To return then to the question of the relationship between human and animal reproductive medicine, it is clear that there is a shared history of technical developments but far less clear whether these practices, as they have come to be socially instantiated, are subject to the same criteria of production. As Charis Thompson (2005, 253) has argued, even the intensely privatized U.S. market in ART services has resisted the kind of mass commodification that has long been standard in agriculture: "In reproductive technology clinics over the last twenty years, there has been considerable standardization as well as innovation. . . . The aim has not been the value-added harnessing of the productive power of the living tissues used, however, but *reproduction* itself. Embryos are tools and raw materials in ART clinics, but they are increasingly rarely *mere* tools. While they remain on the trajectory of a possible future pregnancy in infertility clinics, their reproductive power is rarely if ever reduced to yet another form of labor."

ARTs, then, participate in the general trend to reincorporate the realm of social, sexual, and biological reproduction within the economic sphere—a trend that can be identified as characteristic of the move to post-Fordist workfare states (Bakker 2003). But this translation of reproduction, via ARTs, into marketable commodities has taken place within strict regulatory limits, precisely because it concerns the realm of *human* reproduction. In most cases then, and certainly within the United States, human reproductive medicine has taken the form of the privatized, domestic service, its distinct moments contracted out to other service providers (sperm and egg donors, surrogate mothers, and so on) in the same way that domestic labor and child care are now increasingly available as commercial services.

In the absence of comprehensive federal regulation, there exists a wide and sometimes confusing diversity of state law, clinic guidelines, and legal precedent regarding the status of reproductive services and tissues, with disputes often being resolved on an ad hoc basis in the courts. In general, however, the legal issues surrounding the status of the embryo and its relationship to its progenitors (surrogate or otherwise) have been addressed through a combination of contract and family law. In the case of disputes relating to frozen embryos, courts have appealed to both inheritance and custody law and traditional property rights. As the philosopher of science Jane Maienschein (2003, 151) has remarked: "Frozen eggs are sometimes considered property, other times potential persons whose rights and needs should be questions of custody rather than ownership." Human embryos, in other words, are at most accorded the status of inalienable, familial property but are barred from circulating as freely tradable commodities.

An exception here is the sourcing of human eggs, where the laws of the market generally prevail. And it is where ARTs connect up with the market in eggs that reproductive medicine in general taps into a less familial, more savage kind of reproductive labor. In the sense that egg markets are increasingly drawing on the underpaid, unregulated labor of various female underclasses, the differences between human reproductive medicine and the brute commodification of labor and tissues that prevails in the agricultural industry becomes difficult to maintain. Moreover, the rise of transactional reproductive work demands that we rethink some of the key assumptions of feminist bioethics, displacing the salient questions from the realm of care, dignity, respect, and the liberal ethical contract (of informed consent) to that of labor relations and unequal exchange. This is a move that was first made by feminist theorists of reproductive labor in the context of the first world, Fordist family wage. Now more than ever, it urgently needs to be reconsidered on a transnational level, in the same way that postcolonial feminists have begun to look at the global dynamics of sex work (Kempadoo and Doezema 1998).

But before returning to this question, let's look at the ways in which reproductive and regenerative medicine mobilize the generativity of the fertilized egg. I am interested in the specific understandings of generation that reproductive and regenerative medicine produce, looking in particular at the technical and theoretical consequences of recent developments in stem cell science.

REPRODUCTIVE AND REGENERATIVE MEDICINE: RETHINKING GENERATION

Like most twentieth-century biotechnologies, reproductive medicine works within the general parameters of the Weismannian theory of generation. This is a theory that conceives of the generation of bodies in the manner of vertical transmission, where hereditary information is passed from generation to generation through the germ line, reproducing itself in the mortal bodies of living beings (the soma) while remaining immortal in and of itself. Following the rediscovery of Mendel's laws of recombination in the early twentieth century, the Weismannian theory of generation would come to exercise a decisive influence on genetic theories of heredity, from early transmission genetics to molecular biology. It also left an important legacy in cell theory, where it informed the idea that the body possesses two distinct lineages of cells, somatic and germ cells, which separate and divide their functions in the early stages of development.

In a landmark study on the growth and aging of cells, the American cell biologist Charles Minot (1908) set out the implications of Weismann's "germinal method" for understanding the microprocesses of cell growth, aging, and death. Essential to the Weismannian perspective, he observed, was the idea that the mortality of somatic life would be reflected, at the cellular level, in certain intrinsic limits to the division and differentiation of somatic cells. If we assume that all cells begin in a state of high embryonic plasticity, growth is the process through which the cell differentiates, attains a specialized function, and contributes to the functional organization of the body. Minot thus formulates an inverse relationship between differentiation, organization, and function, on the one hand, and cellular potency, on the other. This relationship would long be considered a rule of cellular development.

According to this rule, all somatic cells, in differentiating, are subject to a loss of potentiality; in taking on specific functions, cells sacrifice some of their embryonic plasticity; cell differentiation thus moves through a progressive, irreversible exhaustion of possibilities, to final cell senescence and death. In Minot's words (1908, 215): "When the cells acquire the additional faculty of passing beyond the simple stage to the more complicated organisation, they lose some of their vitality, some of their power of growth, some of their possibilities of perpetuation; and as the organisation in the process of evolution becomes higher and higher, the necessity for change becomes more and more imperative. But it involves the end. Differentiation leads up, as its inevitable conclusion, to death. Death is the price we are obliged to pay for our organisation, for the differentiation which exists in us." Importantly, Minot (ibid., 205–6) establishes these rules of cellular development by reference to a pathological exception—the indifferent divisibility of the cancerous growth: "The phenomenon of things escaping from inhibitory control and overgrowing is familiar. Such escapes we encounter in tumors, cancers, sarcoma, and various other forms of abnormal growth that occur in the body. They are due to the inherent growth power of cells kept more or less in the young type, which for some reason have got beyond the control of the inhibitory force, the regulatory power which ordinarily keeps them in." The cancerous cell, in a word, is one that avoids aging and death by refusing to differentiate. As such, it represents the ultimate, pathological countervalue to the normative rules of Weismannian generation—and can thus only be defined in negative terms: by its indifference to the normal limits to differentiation and division.

But perhaps a fuller understanding of the taxonomy of growth in late-nineteenth- and twentieth-century biomedicine can only be gleaned by looking at the affective associations of the normal and the pathological. In an essay published in 1929 called "Disgust," the phenomenologist Aurel Kolnai ([1929] 2004, 72) argued that biological theories of the pathological—and the feelings of disgust, repulsion, or horror they generate—are based on an implicit sense of the (distorted) relationship between life and death, the living and the dead: "We can draw the conclusion that disgust is provoked by the proximity or by the challenging or disturbing effect of certain formations which are constituted in such a way that they refer in a determinate manner to life and death." What provokes disgust and thus signifies the pathological, he claims, is not so much the absence or negative of life—the lifeless corpse—as the manic, uninhibited *overproduction* of life, life that reproduces itself outside the proper ends of germinal, sexual reproduction and organic form. It is the "surplus of life," "extreme propagation and growth"—not the growth that grows toward death—that we associate with the pathological (ibid., 72, 75).

Thus the specific sickness of cancer lies in the metastasizing overproduction of life rather than its simple negation. If cancer kills, it is not so much through a direct decomposition of the organism, as an extortion of the vital life force of organic life (cellular division), which it deflects from all ends—other than its own accumulation. There is overproduction of life, writes Kolnai, when the generative processes of growth, reproduction, and regeneration escape the boundaries of organic space and time. Its surplus of life "endeavours to break altogether through any boundaries which may be set upon it and to permeate its surroundings" and "exceeds the limits of a real, or as it were quasi-'personal,' purposeful organic unity" (ibid., 73, 62). It is possessed of a "pretentious inflatedness that is according to its intention unlimited" (ibid., 73).

Above all, what defines the pathological growth is its tendency to absolve itself from the dialectic of life and death, to obliterate the internal negative and limit of Hegelian life. In this way, argues Kolnai, the dominant nineteenth-century definition of the pathological implies a monism of life that is wholly resistant to the mathematics of the dialectic.[4] The cancerous growth refuses to submit to the limits of generational time and death and instead pursues its own relentless self-accumulation. Ultimately, however, while Kolnai concedes that the pathological production of surplus does in fact constitute a kind of life, he concludes that it is life devoid of any true productive power. For Kolnai, as for twentieth-century biomedicine in general, the lifelikeness of cancer is essentially sterile. The overproduction of life is always in the last instance fatal: "In this surplus of life there resides non-life, death. That passing into death through the cumulation of life has a character which is particularly distorted in comparison with simply dying or ceasing to exist" (ibid., 74).

When situated within this taxonomy of the pathological, the challenges posed by developments in stem cell science become strikingly apparent. In effect, what biologists claim to have discovered in the ES cell line are many of the properties that until recently would have been associated with cancerous growth. Indefinitely divisible or "immortal," to use the biological term, uniquely plastic in its possibilities of differentiation and even perhaps metastatic, the ES cell line seems to behave quasi-cancerously.[5] But whereas cancer, since the birth of cellular pathology in the mid-nineteenth century, has been defined as the most pathological of growths, contemporary science describes the stem cell as the most benign, regenerative, and therapeutic of cells. And while nineteenth- and twentieth-century theories of the cell considered the "life" of the cancerous cell to be inherently sterile and hence incapable of performing any *real work* in the generation of the body, stem cell science seems to suggest that the quasi-cancerous properties of the ES cell line are in fact enormously productive. Indeed, with recent experiments having succeeded in producing germ cells from cultured ES cells, it is becoming plausible that "life itself" might be more comprehensively defined by the proliferative, self-regenerative powers of the ES cell rather than the Weismannian theory of the germ line.

Here we are reminded of another thread in the genealogy of stem cell research, one that traces its origins back further than reproductive medicine to earlier work on cellular regeneration, cancer, and embryogenesis. Of particular importance here are a series of studies carried out on embryonal tumors in the 1960s and 1970s, where the term "embryonic stem (ES) cell" was first used interchangeably with that of "embryonal carcinoma (EC) cell." These embryo-like tumors (teratomas or teratocarcinomas) were of interest to cellular biologists because they seemed to caricature the normal processes of embryogenesis, challenging the prevailing definitions of normal and pathological growth.[6]

Thus, even while reproductive and regenerative medicine draw on the same repertoire of experimental tissues, they put them to work in very different ways. For reproductive medicine, as Charis Thompson has observed, the whole point is to culture the fertilized egg cell to term—in other words, to actualize its biological promise in the form of the future individual organism. In this way, even when reproductive medicine industrializes the processes of Weismannian reproduction—freezing and dissecting its constitutive moments into exchangeable time equivalents, standardizing the production of bodies—it does not question its essential premises. In contrast, what stem cell science seeks to produce is not the potential organism—nor even this or that particular type of differentiated cell—but rather *biological promise itself, in a state of nascent transformability*. More precisely, it seeks to discover the culture conditions under which the biological promise becomes *self-regenerative, self-accumulative, and self-renewing*. It wants to culture the ES cell in such a way that it is able to perpetually regenerate its own potentiality, in the form of a not-yet realized surplus of life.

SELLING GENERATION: FROM FAMILY CONTRACTS TO THE EMBRYONIC FUTURES MARKET

The differences between reproductive and regenerative medicine are not merely technical, however. In the United States both sectors are characterized by a remarkable absence of federal regulation (a situation that contrasts starkly with that of the United Kingdom or Germany, for example). But in all other respects, the North American enterprise of stem cell science has been incorporated into different economic infrastructures to those that predominate in reproductive medicine.

The field of ART is characterized by a complex landscape of legal relations, where the outright commodification of certain tissues (eggs, for example) coexists with highly restricted forms of commodification (in the case of embryos), along with other commercial constructs such as contracts for services (clinical tests, procedures, and surrogacy agreements) and hybrid legal forms of custodial and property right (again, where the embryo is concerned). At one end of the spectrum the legal language surrounding ARTs seems to be pulling in the direction of inheritance law, while at the other there is a push toward the kinds of mass commodification that have long prevailed in the business of industrialized agriculture. In contrast, stem cell science in the United States has been subject to remarkably different norms of commercialization and legal valuation. Here, what has prevailed is not so much the commodification of tissues and processes—or a limited form thereof—but rather their integration into the highly financialized, promissory forms of accumulation that I analyzed in chapter 1. What is being constituted here, I suggest, is something like *a market in embryonic futures*, one that brings the promise of capital together with the biological potentiality of cell lines and attempts to conflate the two. The development of both markets and technologies specializing in biological promise can in turn be understood as part of a larger trend toward the intensification of futures trading in both the U.S. and world capital markets.

Futures markets in agricultural commodities have operated in the United States for more than a hundred years and have long included a range of standardized speculative instruments, from futures proper to options and forward contracts. These forward-looking instruments—known as derivatives—were originally designed as a means of hedging against the risk of unexpected price changes, although they have more often than not served the purposes of speculation. As a result of the deregulation of banking and financial markets in the 1970s, derivatives markets, derivative instruments, and trading volume underwent an extraordinary expansion. Futures markets in commodities now exist throughout the world and extend to commodities other than the proverbial pork bellies. In the meantime the U.S. futures market has gradually moved from selling futures on commodities to selling futures on cash, and finally to selling futures on futures contracts. As cultural critic Mark C. Taylor (2004, 167) has written: "As trading things or options on things increasingly gave way to exchanging money and buying and selling intangible options and futures on currencies, the nature of the game changed. Exchanges began to resemble high-stake casinos more than agricultural markets. . . . With the shift from betting on stuff to first betting on stocks, then on indexes, options and futures, gambling turns back on itself and investing becomes the post-modern game of betting on bets."

Financial markets, in other words, have moved from selling futures on tangible things to accumulating promise from promise, through the trading of self-referential futures contracts. Although at one level this might be taken to signify the ultimate triumph of the sign and a delirious abstraction from the tangible world of commodities, it is clearly also much more than that—for in a real sense the post-Fordist model of accumulation brings speculation into the very core of production, so that the two become inseparable.[7] This has been particularly true for emerging, high-risk areas of life science experimentation such as stem cell science, where initial public offerings on the capital markets are enabled through early investments by venture capital funds (many of which are owned by the large mutual and pension funds). There is thus a tight institutional alliance between the arts of speculative promise and risk-taking and the actual cultures of life science experimentation.[8] Indeed, it seems to me that we are seeing a mutual exchange between epistemic, experimental, and commercial modes of speculation such that the recent life sciences are increasingly attuned to the indeterminate promises of cellular life itself. (The question then becomes—to what extent is science as a practice able to demarcate itself from the commercial imperatives of its funding?)

The burgeoning U.S. stem cell market is one instance in which the logic of speculative accumulation—the production of promise from promise—comes together with the particular generativity of the immortalized embryonic stem cell line, an experimental life form that also promises to regenerate its own potential for surplus, indefinitely. It might even be argued that the contemporary form of capitalist accumulation is one in which the permanently nascent dynamic of financial capital attempts to materialize itself, in extremis, in the self-regenerative, embryoid bodies of stem cell science. What Marx referred to as the "automatic fetish" of financial capital here attempts to engender itself as a body in permanent embryogenesis. This in any case has been the quite explicit strategy of Geron, a biotech start-up whose history in many ways exemplifies the speculative turn of life science commercialization.

POST-GERONTOLOGY

Founded in 1990 and publicly traded on NASDAQ since 1996, Geron exemplifies many other biotech start-ups that emerged in the 1990s, financed by venture capital and offering little more than the hope of future revenues from its patent portfolio. Geron is one of the companies that has most successfully maneuvered itself to monopolize the emerging field of regenerative medicine—if and when it starts to return profits. As its name would suggest, Geron specializes in all aspects of cellular aging and exploits both the diagnostic and therapeutic, the pathological and regenerative potential of cellular immortalization: from targeting the immortalization of cells associated with cancer to exploiting the same properties of stem cells as an inexhaustible source of cell production. Its three major product platforms include diagnostic and therapeutic compounds for targeting telomerase (the enzyme that confers replicative immortality to cancerous cells); methods for deriving, maintaining, and scaling up undifferentiated human ES cell lines; and differentiating them into therapeutically relevant cells and cloning technologies.[9] In the absence of federal funding support in the United States, it was Geron that sponsored the pioneering stem cell research carried out by researchers James Thomson at the University of Wisconsin and John Gearhart at Johns Hopkins University. In a strategic move to combine ES cell research with cloning, in 1999 Geron purchased the cloning division of the Roslin Institute (of Dolly fame) and launched a research program with the aim of producing immuno-compatible transplant tissues.

As a presage of its future growth prospects, Geron's *Annual Report* for 2000 describes the miraculous regenerative properties of the company's most celebrated potential product, the human ES cell line: "Unlike all other stem cells discovered to date in humans, human embryonic stem cells (hES cells) can develop into any of the body's cells, including heart, muscle, liver, neural and bone cells. These cells are also unique in that they express telomerase at a high constant level, enabling them to repopulate themselves in an undifferentiated state. Due to this ability for self-renewal, hES cells are a potential source for the manufacture of all cells and tissues of the body. . . . Geron and its collaborators have shown that hES cell clones retain indefinite replicative capacity as well as the ability to form all types of cells in the body" (Geron 2000).

In the literal form of the self-regenerative stem cell line, Geron seems to be holding out a solution to the ailing growth prospects of the pharmaceutical sector (one of the most important potential investors in the field). If the blockbuster drugs of the pharmaceutical industry are facing imminent obsolescence, regenerative medicine promises to give it a new lease on life—a postgerontological fix. And yet the future commercial prospects of this product remain inherently uncertain. In an annual report on its financial condition issued in 2003,

Geron offers a disclaimer on its attempts to predict the unpredictable in the following helpful terms: "This Form . . . contains forward-looking statements that involve risks and uncertainties. We use words such as 'anticipate,' 'believe,' 'plan,' 'expect,' 'future,' 'intend' and similar expressions to identify forward-looking statements. These statements appear throughout the Form . . . and are statements regarding our intent, belief, or current expectations, primarily with respect to our operations and related industry developments. You should not place undue reliance on these forward-looking statements. . . . Our actual results could differ materially from those anticipated" (Geron 2003b).

In much the same way that the undifferentiated stem cell line remains capable of multiple, pluripotent futures, the future returns on regenerative medicine cannot be foretold. As of 2003, most of Geron's therapeutic products have remained in the first stages of clinical trial and have yet to pass the lengthy product development process required before being approved for commercialization. As new technologies, these products run the risk of unforeseen side effects, for which Geron has no product liability insurance (the risk is essentially incalculable).[10] In 2003, Geron (2003b) reported that it had never made a profit on its therapeutic products and does not expect to "for a period of years, if at all." Since going public, its stock price valuation has fluctuated wildly in response to press releases, pro-life reaction, and presidential announcements. As an industry report recently stated, the field of regenerative medicine as a whole has not yet secured the faith of investors, while coveted collaborations with the pharmaceutical sector remain elusive (see Fletcher 2001, 204–5).

Faced with this horizon of fundamental uncertainty, the only thing sustaining Geron's faith in the future is its claim on the intellectual property rights of all future inventions using its products. The human ES cell line seems to prove that "life itself" is capable of overcoming all limits to growth, but of course its actual economic value lies not so much in its powers of self-regeneration as in the formulation of an essentially new form of property right designed to capture its future possibilities of growth—*even when they defy all prediction.*[11] The biological patent responds to the unpredictable potentiality of the ES cell line by inventing a property right over the uncertain future. Under the extremely generous provisions of the U.S. Patent and Trade Office, Geron holds exclusive rights to develop the unmodified ES cell lines isolated by Dr. James Thomson into three modified cell lines for commercial purposes. These three differentiated stem cell lines happen to coincide with some of the most important medical applications.[12] The patent cannot guarantee that the promise of regenerative medicine will be realized. However, it does ensure that if it is actualized, and whatever form its future inventions take, the proceeds will be going to Geron. Life at its most unpredictable will have been precapitalized. The self-regenerative life of the ES cell line—if its promise is realized at all—will emerge in the guise of surplus value.

The decisive political importance of patent rights cannot be overstated here—the biotech "revolution" would have been inconceivable without a full-scale legislative and political campaign to revolutionize property itself, beyond the paradigm of persons and things and beyond the limits of the industrial patent. Below I consider the ways in which recent developments in patent law have overhauled the legal status of both technical invention and biological regeneration.

REINVENTING GENERATION

Dating back to the era of mechanical and industrial invention, the relatively recent tradition of patent law reflects a fundamental division of labor between technical production and biological life. Between the seventeenth and the late nineteenth centuries patent law defined the act of invention as a transformation of inorganic matter, thus excluding both the laws of nature and biological reproduction of all kinds (plant, animal, human) from the sphere of the *machinic* in general. The legal historian Bernard Edelman has pointed out that up until this point, patent law was able to coexist in harmony with the much older tradition of legal personhood (the person as a legal and spiritual entity, a subject of right defined in opposition to the material thing or object of right). Until the late nineteenth century and with few exceptions right up until the 1980s, the legal discourse on invention and the industrial model of technology found a point of convergence in the idea that the "person" should remain outside of the realm of exchange relations; that human generation could not be conflated with the processes of economic reproduction or included within the laws of invention (B. Edelman 1989, 969–70).

In its strict separation of spheres the industrial revolution presupposed a mutually exclusive, yet *analogical* relation between biological reproduction, the transmission of property, and the laws of invention. European and U.S. patent law, for example, describe invention as an act of original "conception," while figuring the "reproduction" of the patented machine as a transmission of the father's name (Strathern 1999, 163–65). In a similar way the early twentieth-century science of genetic heredity borrowed its language from the laws of inheritance: when Weismann formulated the theory of germinal reproduction at the turn of the century, he conceived of life itself by analogy with the transmission of legal personhood (the germ line, he claimed, was equivalent to a "person" and left "in trust" to each generation).

At the beginning of the twenty-first century, both British and U.S. patent law continued to exclude human reproduction from the sphere of invention. In the United Kingdom, for example, human embryonic stem cells defined as "totipotent" could not be patented since they had "the potential to develop into an entire human body" (British patent law here relied on a distinction between totipotent and pluripotent cells, which was far from being established scientifically).[13] In the United States the exclusion of the person from the realm of patentable invention has only recently been enshrined in law. In 1987, Donald J. Quigg, the assistant secretary of commerce and commissioner of patents and trademarks, issued a memo stating that "a claim directed to or including within its scope a human being will not be considered to be patentable subject matter" (quoted in Slater 2002). In 2004 the U.S. Congress made this official when it prohibited the U.S. Patent and Trade Office from issuing patents on human organisms: according to the wording of the bill, "none of the funds appropriated or otherwise made available under this Act may be used to issue patents on claims directed to or encompassing a human organism" (quoted in Wilkie 2004, 42).[14] Such a statement raises the obvious question of how to define the human organism, and what should be construed by the phrase "to encompass a human organism." As one journalist has pointed out, never before have intellectual property lawyers been so preoccupied by the ontological problem of our humanness (Slater 2002).

The problem, however, lies not so much in the legal tradition of patent law itself (which is quite clear about the immunity of the human from commercial relations), but rather in the growing tension between biological and legal discourses on human generation. While bioethics councils continue to rehearse the traditional defense of human dignity ad nauseam, stem cell science is beginning to reformulate the whole problematic of mammalian regeneration in such a way that it undercuts both the legal concept of personhood and the Weismannian model of germ-line reproduction.[15] Regenerative medicine is not so much interested in the reproduction of the human person as in the stem cell's capacity for multiple differentiation and indefinite renewal, in excess of the developmental limits of human morphology. If biologists are increasingly defining the embryonic stem cell as a kind of proto-life, the initial and ultimate source of all human (re)generation, they also distinguish it from the reproduction of the germ line and the transmission of personal humanness. As one stem cell specialist has written, the embryonic stem cell line is "not equivalent" to the potential person in its powers of development (Pederson 1999, 47). It is therefore patentable, according to existing legal definitions of invention.

Accordingly the U.S. Patent and Trade Office is becoming ever more strident in its insistence that the person should remain outside commercial exchange, while in the meantime it is quite logically able to affirm that human life can be reinvented, regenerated, revalorized, *as long as it leaves the biological person intact.* It was on these grounds that in 2001 the U.S. Patent and Trade Office issued patent number 6,200,806 over "a purified preparation of pluripotent human embryonic stem cells" as well as the method used to isolate it—a patent that is so extensive as to cover all human embryonic stem cells and the cells that derive from them.[16] The historical importance of this decision cannot be overestimated. For the first time a generic process of human (re)generation, common to all bodies, has been incorporated into the laws of invention—even while it spares the figure of the person. What is at stake here is a profound legal reconfiguration of the value of human biological life. The potential person will not be commodified—but the surplus life of the immortalized human stem cell will enter into the circuits of patentable invention. This recent extension of patent law equates the self-regeneration of the ES cell with the accumulation of surplus value: as the cell line is subdivided, expanded, and circulated among researchers, its "intellectual value" accumulates and multiplies, returning to the patent holder in the form of interest. It is this property right that decides, through the force of law, that the self-regeneration of life will coincide with the self-valorization of value, that the future materializations of the stem cell will have been appropriated even before their birth into a determinate form.

COMMODIFICATION OR FINANCIALIZATION?

How should we define the particular form of commercialization at work in the biological patent? Are we dealing with the generalized commodification of biological life, as argued by medical anthropologists Nancy Scheper-Hughes and Loïc Wacquant (2002), among others, or rather with an entirely different kind of economic valuation? What is at issue, I suggest, is not simply the suspension and equalization of reproductive time, but a much more radical transmutation of life's value and productivity. When patent law apprehends the value of the stem cell line, it is not in the first instance as an exchangeable equivalent (Marx's definition of the commodity) but as a self-regenerative surplus value, a biological promise whose future self-valorizations cannot be predetermined or calculated in advance. In this way it redefines the value of life as self-accumulative, both on a material level (cell line technology as a deliberate cultivation of the self-regenerative potentialities of living tissue) and a commercial level (the intellectual value of the cell line is not pregiven in the cell line in its unicity but accumulates and multiplies as the cell line is subdivided, expanded, and circulated among researchers).[17]

Accordingly, what we should be looking to here is not so much Marx's reflections on the fetishized commodity form but rather his formula for capital, where he is interested in the peculiar generativity of financial or interest-bearing capital.[18] Financial capital, Marx points out, may index the singular thing or commodity but is never equivalent to it—trading futures for futures in order to generate surplus, its speculative value can no longer be said to reference any kind of fundamental substance, outside of exchange.[19] Presciently, Marx describes the self-valorizing logic of financial capital as a kind of self-regenerative life, somewhere between the miraculous and the monstrous: "Value is here the subject of a process in which, while constantly assuming the form in turn of money and commodities, it changes its own magnitude, throws off surplus-value from itself considered as original value, and thus valorizes itself independently. For the movement in the course of which it adds surplus-value is its own movement, its valorization is therefore self-valorization. . . . By virtue of being value, it has acquired the occult ability to add value to itself. It brings forth living offspring, or at least lays golden eggs" (Marx [1867] 1990, 255).

In short, what is at stake and what is new in the contemporary biosciences is not so much the commodification of biological life—this is a foregone conclusion—but rather its transmutation into speculative surplus value.[20] This is, in the first instance, an actual material process. Immortalized cell lines, quasi-cancerous growths, parthenogenetic conceptions—the life sciences are interested in culturing "life itself" in a state of permanent embryogenesis.[21] However, there is nothing that predestines such promissory life forms to commercial ends. The transmutation of self-regenerative life into self-valorizing capital requires a decisive legislative maneuver. As a retroactive effect of the legal decision to patent the cell line (and the act of recognizing a property right always implies the use of political force), the biological promise is recaptured within the form of an economic surplus value, "value which is greater than itself" (Marx [1867] 1990, 257).

It is not a question here of maligning the pathological excesses of financial capital in favor of the legitimate value-creating powers of industrial production or the measured movements of commodity exchange. The idea that financial capital is essentially cancerous—and thus an illegitimate extortion of the fundamentals of production—is the hallmark of moralizing responses to capital.[22] What I am attempting instead is to question the idea of a proper or organic (re)production of life and labor, while resituating the critique of political economy on a terrain where even the most "pathological" of growths can be productive.

GLOBAL EGG MARKETS

As of yet and with few exceptions, the market in stem cells and related products remains a research market, in which transactions are confined to the exchange and sale of tissues, patents, and knowledge between laboratories. For this reason any analysis of the possible futures of the market in regenerative medicine remains hypothetical.[23] But what of the bodies whose clinical, reproductive, or biomedical labor will be called upon to generate the body of capital, if the market in regenerative medicine is ever successfully commercialized? The development of commercial egg markets, in both the United States and Eastern Europe, as a way of supplementing the endemic shortage of eggs available for IVF procedures, provides an insight into the possible future form of an integrated bioeconomy. Here, it seems, we are seeing the emergence of a new market in clinical reproductive labor, one that is developing in close synergy with preexisting transnational economies of feminized labor (domestic, sexual, and maternal).[24] Recent exposés of the burgeoning egg markets in Eastern Europe make it clear that women who participate in any one sector of this reproductive economy are likely to migrate to another, so that the boundaries between actual biomedical, reproductive labor on the one hand and sexual and domestic labor on the other are extremely fluid. With the trend to outsource clinical trials to such countries as India and China, it seems likely that tissue sourcing and tissue markets may follow the same trajectory.

At the moment these trends are so recent that it is difficult to make any forecasts about where they are going.[25] But at the very least they point to the kinds of hard, clinical labor that a "successful" transnational market in biomedical services is likely to generate. What comes to light here is the violence of the debt form and the flipside of the promise—if it is to actualize at all, capital's future embryoid body will need to draw on a continuous "gifting" of reproductive labor and tissues. In this way capital's dream of promissory self-regeneration finds its counterpart in a form of directly embodied debt peonage. What embryoid capital demands is a self-regenerating, inexhaustible, quietly sacrificial source of reproductive labor—a kind of global feminine. Its mystification lies in the belief that the embryoid body is capable of regenerating itself.

Again, although it is too early to predict the global dynamics of this emerging reproductive economy, it is already possible to establish a number of points of difference and comparison with the reproductive politics of welfare state and colonialist biopolitics. The postmodern mode of bioproduction is not eugenicist; it is not interested in establishing the norms of standardized reproduction but rather in effecting a generalized destandardization of bodily generation. This difference is expressed in the current confusion around ideas of normal and pathological growth, both in the life sciences and economics. Moreover, the neoliberal administration of reproductive life is not nation-based. On the contrary, what is increasingly visible in the realm of reproductive services is a progressive privatization and denationalization of feminized, maternal labor. Summarizing these trends, it could be argued that neoliberal biopolitics abandons the ideal of reproductive labor and the family wage as a national biological reserve—a compulsory "gift" of life in the service of the nation—and transfers its promise into a speculative future, where the technological capabilities of the biotech revolution are credited with overcoming all limits to growth in the present.

However, this does not mean that the identitarian appeal to the fundamentals of (re)production and the inherent value of life have been abandoned. On the contrary, the past few decades have witnessed an extraordinary proliferation of neofundamentalist movements, intent on reestablishing the fundamentals of ethnic, national, and racial identity even in the face of a hopelessly uncertain, transnational future. In the next chapter I am particularly interested in the rise of the North American evangelical movement, the neofundamentalist movement that speaks out most stridently on the issues of life and reproduction. It is highly significant that this movement has taken up the basic tenet of mid-century human rights discourse—the right to life—and reformulated it in a future mode, as the right to life of the potential or unborn person. This act of renaming has been so successful that the pro-life meaning of the "right to life" has largely effaced its original context and import. In so doing, it points to a defining characteristic of neofundamentalist movements today—their tendency to project the question of property, nationhood, and obligation into the future, as a horizon of prophetic restoration.

◻ 6 ◻

THE UNBORN BORN AGAIN

Neo-Imperialism, the Evangelical Right, and the Culture of Life

> I also believe human life is a sacred gift from our Creator. I worry about a culture that devalues life, and believe as your President I have an important obligation to foster and encourage respect for life in America and throughout the world.
>
> —George W. Bush, 2001

IN EARLY 2002, GEORGE W. BUSH ISSUED A PRESS RELEASE PROCLAIMING January 22 as National Sanctity of Human Life Day (White House 2002). In the speech he delivered for the occasion, Bush reminded the public that the American nation was founded on certain inalienable rights, chief among them being the right to life. The speech is remarkable in that it assiduously duplicates the phrasing of popular pro-life rhetoric: the visionaries who signed the Declaration of Independence had recognized that all were endowed with a fundamental dignity by virtue of their mere biological existence. This fundamental and inalienable right to life, Bush insisted, should be extended to the most innocent and defenseless among us—including the unborn: "Unborn children should be welcomed in life and protected in law" (ibid.).

What is even more remarkable about the speech is its smooth transition from right to life to neoconservative "just war" rhetoric. Immediately after his invocation of the unborn, Bush recalls the events of September 11, 2001, which he interprets as an act of violence against life itself. These events, he claims, have engaged the American people in a war of indefinite duration, a war "to preserve and protect life itself," and hence the founding values of the nation. In an interesting confusion of tenses, the unborn emerge from Bush's speech as the innocent victims of a prospective act of terrorism while the historical legacy of the nation's founding fathers is catapulted into the potential life of its future generations. Bush's plea for life is both a requiem and a call to arms: formulated in a nostalgic future tense, he calls upon the American people to protect the future life of the unborn in the face of our "uncertain times" while preemptively mourning their loss.[1]

In the wake of the terrorist attacks of September 11, 2001, it is easy to forget that the most explosive test confronting Bush in the early months of his presidency was not terrorism but the issue of whether to provide federal funds for research on embryonic stem cells. The issue had been on the agenda since 1998, when scientists funded by the private company Geron announced the creation of the first immortalized cell lines using cells from a frozen embryo and an aborted fetus. Bush, who had campaigned on an uncompromising pro-life agenda, put off making a decision for as long as possible. In July 2001 he made a visit to the Pope, who reiterated the Catholic Church's opposition to any experimentation using human embryos (White House 2001). On August 11, 2001, however, Bush made a surprise announcement, declaring that he would allow federal funding on research using the sixty or so embryonic stem cell lines that were already available (the actual number of viable cell lines turned out to be fewer than this). In making this concession to stem cell research, he claimed, the U.S. government was not condoning the destruction of the unborn. "Life and death decisions" had already been taken by scientists, Bush argued. By intervening after the fact, the state was ensuring that life would nevertheless be promoted, in this case not the life of the potential person but the utopia of perpetually renewed life promised by stem cell research.

In the months leading up to his decision, Bush had attempted to soften the blow for the religious right by extending universal health coverage to the unborn, who thereby became the first and only demographic in the United States to benefit from guaranteed and unconditional health care, at least until the moment of birth (Borger 2001). However it translates in terms of actual health care practice, the gesture was momentous in that it formally acknowledged the unborn fetus as the abstract and universal subject of human rights—something the pro-life movement had been trying to do for decades.

In the meantime and in stark contrast to the U.S. government's official moral stance on the field of stem cell research, U.S. legislation provides for the most liberal of interpretations of patent law, allowing the patenting of unmodified embryonic stem cell lines. For this reason the most immediate effect of Bush's decision to limit the number of stem cell lines approved for research was to ensure an enormous captive market for the handful of companies holding patents on viable stem cell lines. One company, in particular, is poised to profit from Bush's post life and death decision. The aptly named Geron, a start-up biotech company specializing in regenerative medicine, also happens to hold exclusive licensing rights to all the most medically important stem cell lines currently available. Uncomfortably positioned between the neoliberal interests of the biomedical sector and the religious right, Bush seems to have pulled off a political tour de force: while proclaiming his belief in the "fundamental value and sanctity of human life," he was also able to "promote vital medical research" and less ostentatiously to protect the still largely speculative value of the emerging U.S. biotech sector.

In his press release announcing the new National Sanctity of Human Life Day, Bush was expressing his faith in the future of life. But what kind of future does he believe in? And what tense is he speaking in? Bush's pro-life rhetoric oscillates between two very different visions of life's biomedical and political future: one that would equate "life itself" with the future of the nation, bringing the unborn under the absolute custodial protection of the state and the family, and the other that less conspicuously abandons biomedical research to the uncertain and speculative future of financial capital investment. On the one hand, life appears as an inalienable gift, one that must be protected at all costs from the laws of the market; on the other hand, the patented embryonic stem cell line seems to function like an endlessly renewable gift—a self-regenerative life that is also a self-valorizing capital.

What appears to be at stake, behind the scenes of George Bush's speech, is the determination of the value of life. How is the promise of biological life to be evaluated? Is its value relative or absolute? Perhaps what is most seriously at issue is the temporal evaluation of life, its relation to futurity (predetermined or speculative). How will this value, whatever it consists of, be realized? Given that the contemporary life sciences are tending to uncover a "proto-life" defined by its indifference to the limits of organic form, within what limits will its actualization nevertheless be constrained? What is interesting about Bush's decision on stem cells is that he comes up with two solutions, whose apparently conflicting appraisals of the value of life cofunction quite nicely in practice. According to media reports, Bush stacked his ethics committees with a half and half mix of pro-life supporters, hell-bent on protecting the sanctity of life, and representatives of the private biomedical sector, just as fervently opposed to any kind of federal regulation of stem cell research. Somehow the two positions managed to coexist in the person of George W. Bush.

In keeping with the general tone of his public declarations, Bush's speeches on the unborn weave together a subtle mix of three tendencies in American political life—neoconservatism, neoliberal economics, and pro-life or culture of life politics. These three tendencies have coexisted in various states of tension and alliance since the mid-1970s. But they have been getting closer. Such neoliberals as George Gilder have started to openly affirm their evangelical faith. Such neoconservatives as William Kristol have aligned themselves with the evangelical right in its defense of the right to life and its opposition to stem cell research. Both have more recently championed the cause of creationism in American schools. Michael Novak, the free-market Catholic neoconservative, has always quite happily embodied the tension between a capitalism of endless growth and an unshakable faith in the absolute limits of life. In the meantime evangelicals who were once content to fight over domestic moral and racial politics have embraced an increasingly militant and interventionist line on U.S. imperialism, seeing U.S. victory in the Middle East as the necessary prelude to the End Times and the second coming of Christ. Under Bush as president and indeed in the person of Bush, these tendencies have become increasingly difficult to distinguish.

The biography of George W. Bush is in many ways in keeping with the profound transformations that have taken place in U.S. Protestantism over the past three decades. Brought up as a mainstream Methodist, Bush was born again as an evangelical Christian around age forty (Kaplan 2004, 68–71; K. Phillips 2004, 229–44). In the process he moved from a religion based on personal self-transformation and discipline to one that espouses a decidedly more expansive, even world-transforming philosophy. More than one of Bush's close associates have commented that he saw his investiture as president of the United States as a sign of divine election, one that linked his personal revival to that of America—and ultimately to that of the world. Such luminaries of the evangelical right as Pat Robertson could only agree with Bush. After all, it was largely thanks to the (white) evangelical right that he had won the 2000 elections (Kaplan 2004, 3). And in return, the Bush administration allowed them an unprecedented influence in almost all areas of government policy (ibid., 2–7).

Bush's economic philosophy, too, reflects a dramatic transformation in

Protestant views on wealth and sin. The ethic of late Protestantism is much more investment- than work-oriented, much more amenable to the temptations of financial capital than the disciplines of labor, and evangelical Christians have found a welcome ally in the writings of various free-market and supply-side economists. In his 2004 biography of the Bush family clan, Kevin Phillips has convincingly argued that George W. Bush himself is also essentially a supply-sider: despite appearances, his economic outlook is in fact much more informed by his experience in investment banking and finance (think Enron) rather than the nuts and bolts of the oil industry.

Bush's conversion to the neoconservative cause was less immediate, and perhaps more contingent on the events of September 11, than is commonly recognized. In their careful study of the Bush team's defense policy prior to late 2001, the political theorists Stefan Halper and Jonathan Clarke (2004, 112–56) have pointed out that the early Bush was notably reluctant to engage in any gratuitous nation building. But the reasons for such an alliance, when it did happen, were certainly not lacking—since the mid-1970s, the neoconservatives had strategically aligned themselves with the prophets of supply-side economics, and during the 1990s their attentions turned to the populist appeal of the right-to-life movement (ibid., 42, 196–200). In the aftermath of the September 11 attacks, they were able to present George W. Bush with a ready-made blueprint for war, one that would satisfy both the millenarian longings of the Christian right and the evangelistic tendencies of free-market capitalism.

How have these imperialist, economic, and moral philosophies been able to work so tightly together under Bush's presidency, and why have they converged so obsessively around the "culture" of promissory or unborn life? To address these questions, I first look at Georg Simmel's 1907 work on the relationship between economics and faith. I then turn to a discussion of the links between Protestantism and capitalism and more pertinently between the history of American evangelical revivals and the specific cultures of American liberalism and *life*. Evangelical Protestantism, I suggest, has developed a doctrine of debt, faith, and life that differs in fundamental respects both from the Roman Catholic tradition and mainline, Reformationist Protestantism. It is imperative to have some understanding of these differences in order to grasp the impulses informing the "culture of life" movement today.

It is equally important, however, to look at the ways in which the evangelical movement has itself mutated over the past three decades, reorienting its traditional concerns with life, debt, and faith around the focal point of sexual politics. The neo-evangelical movement, I argue, combines the revolutionary, future-oriented impulse of earlier American revivals with a newfound sexual fundamentalism.[2] It is this contrarian impulse that informs George W. Bush's culture of life politics and is reflected perhaps most forcefully in his ambivalent stance on stem cell research. It is also characteristic of the tendencies of capitalism today, in which a speculative reinvention of life comes together with a violent desire to reimpose the fundamentals, if only in the figure of a future or unborn life.

ECONOMICS AND FAITH

Increasingly, it would seem, it is becoming difficult to confront the most violent manifestations of contemporary economic imperialism without at the same time thinking through their religious, salvationist, and faithlike dimensions. Yet there is little in the contemporary economic literature that would enlighten us on the relationship between the two.[3] One notable early exception is sociologist Georg Simmel's *Philosophy of Money*, a work that combines anthropological, historical, and economic perspectives on the emergence of modern capitalism in ways that might still prove fruitful. Simmel begins by noting that all economic relations, to the extent that they require trust in the future, involve a certain element of faith. Yet it is only in a money economy that this faith goes beyond a simple inductive knowledge about the future and takes on a "quasi-religious" flavor (Simmel 1978, 179). A money economy, after all, is one in which the object to be exchanged (money) is itself born of faith: all money is created out of debt and is therefore of a promissory or fiduciary nature, even before it is exchanged. Simmel draws attention to the two-sidedness of this faith: money on the one hand embodies a promise (to the creditor) and on the other hand a threat of violence (to the debtor); it brings together obligation and trust.

In the case of market economies this two-sided faith relation is extended to all members of a community. A capitalist economy, Simmel asserts, is one in which the whole life of a community is indebted to the debt form. But having established its quasi-religious nature, how does he define the particular religious form of capitalism? What kind of faith does capitalism require? What is the temporality of its promise and its obligation? And what are its specific forms of violence? In Simmel's historical account of capitalism he makes it clear that the emerging market economies of the early modern period fundamentally differed from and disrupted the established forms of sovereign medieval power with their close ties to the Catholic Church and their foundations in landed wealth. A basic premise of his argument is that the philosophy of money needs to be distinguished from the various political theologies of sovereign power. What then is the difference between the philosophy of early modern Christian faith, which we have largely inherited from the Middle Ages, and the quasi-religious faith of capitalism?

It should be noted that the philosophy of Roman Catholicism, as exemplified in the work of someone like Thomas Aquinas, is at one and the same time a political and economic theology, inasmuch as the authority of the medieval church extended to both domains. What unites these spheres, in the work of Aquinas, is a common understanding of foundation, origin, and time (the transcendent or the eternal). This idea of foundation is most clearly enunciated in the doctrine of the Gift, which brings together the questions of theological, political, and economic constitution. In Aquinas's work (1945, 359–362) the Holy Spirit is the Gift of Life that reunites the finite and the infinite incarnations of the Holy Trinity. As such, the Gift is also the originary act through which God creates life, so that from the point of view of his creatures, life is a series of debt installments, a constant quest to repay the wages of sin. Implicit in his theology is the notion that the Gift (which is also a debt) is underwritten by an original presence, the eternal unity of finite and infinite, in which all debt is canceled out. In this way Christianity promises the ultimate redemption of the debt of life, a final reunion of the finite and the infinite, even if it is unattainable in this world. It instructs the faithful to believe in a final limit to the wages of sin.

If we turn to Aquinas's work on jurisprudence, which includes a consideration of price and exchange, it becomes apparent that his economic philosophy shares precisely the same mathematics of debt.[4] His premise here is that any institutionalized political form such as the state must be underwritten by a stable referent or use value, an ultimate guarantor of the value of value, in order to maintain a proper sense of justice. In this way Aquinas's economic philosophy is founded on the possibility of debt redemption. All exchange values must be measurable against a "just price," in the same way that each human life is redeemable against an original Gift.

Historical work on the economic philosophy of the Middle Ages has emphasized just how closely such ideas reflected the actual position of the early Christian Church.[5] The medieval church was an economic and political power in its own right, one whose wealth was based in landed property rather than trade. For this reason the church was not opposed to a certain level of state regulation of exchange and price control, as long as these worked to maintain the "just price" of church property, while it was virulently opposed to certain forms of trading profit, particularly usury. Usury, after all, is a credit/debt relation that wagers on the instability of price. It aims to create money out of a perpetually renewed debt, and it does this without recourse to a fundamental reserve or guarantor of value. It has no faith in the measurability of value and no interest in the final redemption of debt.

It is here that Simmel locates the fundamental difference between the early economic theory of the Christian Church and the particular faith form of modern capitalism. As a form of abstraction, he argues, the capitalist economy dispenses with all absolute foundation, all possibility of final measure, all substantial value: "The fact that the values money is supposed to measure, and the mutual relations that it is supposed to express, are purely psychological makes such stability of measurement as exists in the case of space or weight impossible" (Simmel 1978, 190). By asserting this, of course, Simmel does not want to deny the historical existence of all kinds of institutions designed to uphold the measurability of exchange value (Simmel's *Philosophy of Money* is in part a detailed history of such institutions—from precious metals to the Central Bank to the labor theory of value). Without such institutions and their lawful forms of violence, no creditor would be able to demand repayment. Yet he insists that such institutions, considered singly, are both mutable and not foundational to the creative logic of capitalism. Modern capitalism, in other words, is a social form in which the law no longer figures as a *source* of creation, but rather as an institution charged with the power of sustaining the faith a posteriori, through the threat of violence. In stark contrast to the economic theology of the medieval church, capitalism is a mode of abstraction that generalizes the logic of usury and constantly revolutionizes all institutional limits to its self-reproduction. What then is its particular mode of faith?

BORN-AGAIN NATION: AMERICA, EVANGELICALISM, AND THE CULTURE OF LIFE

This is the question that preoccupies the sociologist Max Weber in *The Protestant Ethic*, where he famously analyzes the historical affinities between the rise of the Protestant faith and the beginnings of modern capitalism. In Calvinism, Weber identifies the first religion to celebrate the life of business and the disciplines of labor, not merely as means to an end but as the very manifestation of faith in God. In stark contrast to the Roman Catholic tradition, with its repudiation of earthly pursuits, Protestantism brings "God within the world" and espouses an immersive, transformative relation to God's creation rather than a contemplative one (Weber [1904–5] 2001, 75). And in late-seventeenth-century variations on Protestantism, argues Weber, there is an even more extreme change in attitudes toward wealth creation. Here even usury, the creation of money out of promise and debt, is accepted as a legitimate way of expressing one's faith. This move away from a strict Calvinist doctrine of predestination, suggests Weber, is reflected most acutely in the rise of later, less "aristocratic" forms of Protestant faith such as Methodism, in which the doctrine of *regeneration* or *the new birth*, as espoused by the evangelical minister John Wesley, becomes central (ibid., 89–90). The Methodist philosophy of conversion through rebirth develops in England but flourishes in America—and it is here that Weber closes his analysis.

At this point, then, Weber's perspective on the European Protestant Reformation needs to be supplemented by an account of the specific inventiveness of American Protestantism—particularly in its understanding of life, faith, and wealth.[6] For a start, as noted by the historian Mark Noll (2002, 5), the most successful currents in American Protestantism were self-consciously evangelical—in other words they practiced a radically democratized form of worship, with a focus on the personal experience of conversion and rebirth. In the process the American take on Methodism not only brought sanctification into this world and this life, but also freed it from the necessity of institutional mediation to an extent that could hardly have been imagined by Wesley himself. For the American evangelicals, being born again was an experience of autonomous although involuntary self-regeneration—the Holy Spirit being wholly implicated in the self, just as the self was implicated in the world.

Moreover, the uniquely American evangelical experience was reflected in an enthusiasm for wealth creation far surpassing its counterparts in the European tradition. Here, suggests Noll (ibid., 174), the anti-authoritarianism of the American evangelicals expresses itself as an aversion to all foundational value, a belief in the powers of money that separates promise from all institutional guarantee and regulating authority, figuring the market itself as a process of radical self-organization and alchemy. In this way the doctrine of the new birth merges imperceptibly with a theology of the free market, one that situates the locus of wealth creation in the pure debt form—the regeneration of money from money and life from life, without final redemption. This is a culture of life-as-surplus that is wholly alien to the Catholic doctrine of the Gift and its attendant political theologies of sovereign power. Pushed to its extreme conclusions, evangelicalism seems to suggest that the instantaneous conversion of the self—which is held to render an ecstatic surplus of emotion—is the emotive equivalent of a financial transmutation of values, the delirious process through which capital seeks to recreate itself as surplus.[7]

The doctrine of regeneration imparts a highly idiosyncratic vitalism to the evangelical understanding of nationhood. Again, as detailed by Noll (ibid., 173–74), the extraordinary rise of Protestant evangelical faith between the Revolution and the Civil War was decisive in fusing together the discourses of republicanism and religious experience, so that in an important sense, the language of American foundation and independence became inseparable from that of evangelical conversion. It is therefore not only in the minds of latter-day fundamentalists that the founding of America came to be figured as an act of God-given grace: such analogies were already sufficiently self-evident in late-nineteenth-century America that Abraham Lincoln was able to refer to Americans as God's almost chosen people, calling for a *new birth* of the American nation itself.

What is the relationship between these earlier forms of American evangelicalism and the right-to-life movement of the 1970s? What has become of the experience of rebirth today? And what are its connections to evangelical views on capitalism? To respond to these questions, we need to look at the ways in which U.S. capitalism itself has mutated over the past three decades, redefining its relationship to the countries of the rest of the world, both creditors and debtors. In what follows, I argue that U.S. imperialism today is founded on the precarious basis of a perpetually renewed debt—and thus seems to take the evangelical doctrine of wealth creation to its extreme conclusions.

This particular form of economic faith is also celebrated in neoliberal theories of wealth creation.

DEBT IMPERIALISM: THE UNITED STATES SINCE 1971

What do I mean by U.S. debt imperialism? In his study of the changing faces of U.S. imperialism, revised and rewritten over three decades, the economist Michael Hudson (2003) has argued that the nature of U.S. imperial power underwent a dramatic change in the early 1970s, when Nixon abandoned the gold-dollar standard of the Bretton Woods era. Hudson was originally hired under the Nixon administration to report on the costs of the Vietnam War and its connection to the budget deficit. In 1972, and at the behest of various federal administrations, he published a full-length book on the question. His conclusions were damning—by demonetizing gold, Hudson argued, the United States had initiated a form of superimperialism, which effectively left the country off the hook in terms of debt repayment. Instead of taking this as an admonition, however, the U.S. administration received it as an unintended recipe for success, one that should henceforth be maintained at all costs. Hudson's book reportedly sold well in Washington, although his work was publicly repudiated. He was promptly hired as an economic adviser at the conservative Hudson Institute.

Hudson's argument is complex and at odds with the mainstream of left-wing commentaries, which tend to see America's spiraling debt as the harbinger of its imminent decline. He identifies the early 1970s as a turning point. Prior to 1971, the United States had acted as a creditor vis-à-vis the rest of the world. After World War II the dollar was convertible against gold and thus remained indexed to a conventional unit of measurement. While the gold standard remained in force, the political and economic limits of the American nation were inherently circumscribed. It was the gold standard that prevented the United States from running up excessive balance-of-payment deficits, since foreign nations could always cash in surplus dollars for gold. As a nation, the United States was underwritten by an at least nominal foundation.

When gold was demonetized, however, the United States abandoned even this conventional guarantor of exchange value. As foreign governments could no longer cash in their surplus dollars for gold, it was now possible for the U.S. government to run up enormous balance-of-payment deficits without ever being held to account. Indeed, it became feasible for the United States, as a net importer, to create debt *without limit* and to sustain its power through this very process. Hudson contends that such a strategy inaugurates a fundamentally new kind of imperialism—a superimperialism that is precisely dependent on the endless issuing of a debt for which there is *no hope of final redemption*. Hudson explains the details of this process as follows: all the dollars that end up in Asian, Eastern, and European central banks as a result of the United States' massive importing now have no place to go but the U.S. Treasury. With the gold option ruled out, foreign nations now have no other "choice" than to use their surplus dollars to buy U.S. Treasury obligations (and to a lesser extent corporate stocks and bonds). This effectively amounts to a forced loan, since in the process what they are doing is lending their surplus dollars back to the U.S. Treasury, thereby financing U.S. government debt.

This forced loan, Hudson (ibid., ix) points out, is a losing proposition, as the falling dollar progressively erodes the value of U.S. Treasury IOUs. It is a "loan" without foreseeable return: U.S. debt can never and will never be repaid; rather, it will be rolled over indefinitely, at least as long as the present balance of international power remains in place (ibid., xv–xvi). The momentum attained by these dynamics is now such that U.S. debt creation effectively functions as the source of world capitalism, the godhead of a cult without redemption. Trends that were initiated in 1972 have now become blatant, particularly under George W. Bush: the U.S. Treasury has run up an international debt of more than $60 billion, a deficit that finances not only its trade but also its federal budget deficit. Moreover, Hudson argues, the cycle of U.S. debt creation has now become so integral to the workings of world trade that the consequences of any upheaval might well appear apocalyptic, even to countries outside the United States.[8]

What does Hudson's work tell us about the character of U.S. nationhood and imperialism today? And how exactly do we define a nation that seeks to recreate itself and world power relations out of a fount of perpetual debt? In terms of traditional theories of economic and political nationhood, Hudson's analysis would seem to lead to the unsettling conclusion that the American state is rigorously devoid of foundation, since the possibility of its continued self-reproduction has come to coincide with the temporality of perpetual debt. As a nation, the United States no longer rests on any minimal reserve or substance but, in synergy with the turnover of debt, exists in a time warp where the future morphs into the past and the past into the future without ever touching down in the present. In economic terms, then, the very idea of the American nation has become purely promissory or fiduciary—it demands faith and promises redemption but refuses to be held to final account. Its growing debt is already renewed just as it comes close to redemption, already born again before it can come to term. America is the unborn born again.

And yet the importance of Hudson's work is to show that there is nothing ethereal about the imperialism of U.S. debt creation. Indeed, it is through the very movement by which it renounces all economic foundation, Hudson claims, that the United States is able to reassert itself as the most belligerent of political forces and the most protectionist of trading partners. The position of the United States at the very vortex of debt imperialism has meant that it has been able to function as a profligate, protectionist state, spending enormous amounts on the military, domestic trade subsidies, and R & D, while the rest of the world's nations have had to subject themselves to the rigors of IMF-imposed budget restraint (ibid., xii). In other words, while the United States, acting through the IMF and the World Bank, imposes the most draconian measures of debt redemption on the rest of the world, it alone "acts uniquely without financial constraint," turning debt into the very source of its power (ibid., xii).

How has the United States ensured that the surplus dollars held by its foreign trading partners would be effectively reinvested in U.S. government securities? According to Hudson, essentially through the use—real or threatened—of institutional violence. The United States exercises unilateral veto power within such purportedly multilateral institutions as the IMF and the World Bank (economists Susan George and Fabrizio Sabelli [1994] have quite seriously analyzed the successive internal reforms of these institutions as so many attempts to establish an orthodox *doctrine of the faith* in the arena of world economic policy). But the economic prescriptions of the World Bank and the IMF have also, necessarily, been backed up by the threat of military retaliation. U.S. diplomats, notes Hudson (2003, ix), have long made it perfectly clear that any return to gold or attempt to buy up U.S. companies would be considered as an act of war. The irony here is that the exorbitant military expenditure of the United States has been financed through the very debt imperialism it is designed to enforce!

All this suggests the need for a highly nuanced interpretation of the nature of U.S. nationalism in the contemporary era, one that takes into account both the deterritorializing and reterritorializing trends of debt imperialism. For it implies that the very loss of foundation is precisely what enables the United States to endlessly refound itself, in the most violent and material of ways. In the era of debt imperialism, nationalism can only be a refoundation of that which is without foundation—a return of the future, within appropriate limits.[9] The endless revolution (rolling over) of debt and the endless restoration of nationhood are inseparably entwined. The one enables the other. And the one perpetuates the other, so that *revolution becomes a project of perpetual restoration and restoration a project of perpetual revolution.* It is only when the double nature of this movement is grasped that we can understand the simultaneously revolutionary and restorative nature of contemporary capitalism in general, its evangelism and its fundamentalism.

U.S. imperialism, in other words, needs to be understood as the extreme, "cultish" form of capital, one that not only sustains itself in a precarious state of perpetually renewed and rolled-over nationhood, but that also, of necessity, seeks to engulf the whole world in its cycle of debt creation.[10] The economic doctrine corresponding to U.S. debt imperialism can be found in several varieties of neoliberalism, in particular the supply-side theories of the Reagan era. Its theological expression can be found in neo-evangelicalism, the various revived and militant forms of Christian evangelical faith that sprang up in the early 1970s. Supply-side economists and neo-evangelicals share a common obsession with debt and creationism. For such supply-side theorists as George Gilder, economics requires an understanding of the operations of faith, and for the right-wing evangelicals who cite him, the creation of life and the creation of money are inseparable as matters of biblical interpretation.

NEOLIBERALISM: THE ECONOMICS OF FAITH

It is surely not coincidental that one of the most influential popularizers of neoliberal economic ideas, the journalist George Gilder, also happens to be a committed evangelical and creationist, whose work argues for the essentially religious nature of economic phenomena.[11] Gilder's classic work, *Wealth and Poverty* (1981), is as much a meditation on the faith as a celebration of U.S. debt imperialism and debt-funded growth. Drawing on anthropological work into the relationship between promise, belief, and debt, Gilder sets out to explain the particular faith form required by contemporary U.S. power. The new capitalism, he asserts, implies a theology of the gift—"the source of the gifts of capitalism is the supply side of the economy"—but one that differs in fundamental respects from Roman Catholic philosophies of debt and redemption (ibid., 28). Here there are no fundamental values, no just price or word against which the fluctuations of faith can be measured and found wanting. Nor is there any final redemption to look forward to. What distinguishes the gift cycle of the new capitalism, claims Gilder, is its aversion to beginnings and ends (ibid., 23).

In the beginning was not the word, God the Father, or even the gold standard, but rather the promise—a promise that comes to us from an unknowable future, like Christ before the resurrection. And in the end is not redemption but rather the imperative to renew the promise, through the perpetual rolling over of U.S. government debt. The promise may well be entirely uncertain, but this does not mean that it will not be realized at all. On the contrary, Gilder insists that it will be realized, over and over again, in the form of a perpetually renascent surplus of life. The return on debt may be unpredictable, but it will return nevertheless (ibid., 25). At least as long as we maintain the faith: "Capitalist production entails faith—in one's neighbors, in one's society, and in the compensatory logic of the cosmos. Search and you shall find, give and you will be given unto, supply creates its own demand" (ibid., 24).

Importantly, what Gilder is proposing here is not merely an economic doctrine but a whole philosophy of life and rebirth. What neoliberalism promises, he insists, is not merely the regeneration of capital but the regeneration of the earth itself, out of the promissory futures of U.S. debt imperialism. It is this belief that informs Gilder's strident antienvironmentalism (and that of many of his evangelical and neoliberal brothers). In a world animated by debt imperialism, there can be no final exhaustion of the earth's resources, no ecological limits to growth that will not at some point, just in time, be renewed and reinvigorated by the perpetual renascence of the debt form itself (ibid., 259–69). His is a doctrine of the faith that not only promises to renew the uncertain future but also to reinfuse matter itself with a surplus of life, over and over again. The irony of this position lies in its proximity to the technological promise of regenerative medicine. The burgeoning U.S. stem cell market is one instance in which the logic of speculative accumulation comes together with the peculiar generativity of the immortalized embryonic stem cell line, an experimental life form that also promises to regenerate its own potential for surplus, without end. What Karl Marx referred to as the "automatic fetish" of financial capital here attempts to engender itself as a body in permanent embryogenesis.

In this way Gilder's theology of capital sustains a belief in the world regenerative, salvationist powers of U.S. debt imperialism. It also offers one of the most comprehensive expositions of the neo-evangelical faith today. It is no coincidence that Gilder's work is frequently cited in the voluminous evangelical literature on financial management, investment, and debt, where the creation of life and the creation of money are treated as strictly analogous questions of theological doctrine.[12] This is a faith that separates the creation of money from all institutional foundations or standards of measurement; a religion that conceives of life as a perpetual renascence of the future, unfettered by origin.

The question of foundation is not overcome, however. On the contrary, Gilder's neoliberal philosophy is exemplary precisely because it brings together the utopian, promissory impulse of speculative capital with the imperative to reimpose the value of value, even in the face of the most evanescent of futures. The problematic can be summarized as follows: How will the endless promise of the debt be realized, distributed, consumed? How are we to restore the foundations of that which is without foundation? How will the gift of capital, which emanates from the U.S. Treasury, be forced to repatriate within the confines of America the nation? After all, it could just as easily not return, go roaming around the world, and reinvest somewhere else—or not at all. Gilder's theology of capitalism is haunted by the possibility that the promissory future of the debt will not be reinvested within the proper limits of the American nation, that the promise that *is* America will not be realized, reborn, rolled over. More generally, perhaps, he expresses the fear that faith in the long run may fail to reinvest in the property form at all—the fear of revolution without restoration, a gift without obligation. The law of value, then, needs to be reasserted; actual limits need to be reimposed on the realization of the future.

For Gilder these limits are of three mutually reinforcing kinds. The first is summed up in the brute law of property: there is no economic growth without inequality, scarcity, and poverty. There is no debt imperialism without debt servitude. The second is of a political kind: economic enterprise must be shored up by a "strong nation," a nation, that is, which has emptied itself as far as possible of all social obligations toward its members, while investing heavily in law and order. Implied in these two conditions are certain limits on the biological reproduction of the American nation: America must continue to reproduce itself as white, within the proper restrictions of the heterosexual family. In this way Gilder's assertion of the law of property is strictly inseparable from his white nationalism and his avowed "moral conservatism." The foundational measure of value *is* the nation, which *is* the property form, which in turn is realized in the most conservative of moral institutions: the straight, white, reproductive family.

It is this amalgam of political, economic, and moral law that gets summed up in the notion of a "right to life" of the unborn. The unborn, after all, is the future American nation in its promissory form, the creative power of debt recontained within a sexual politics of familial life. And as the new right has made clear, its reproduction is the particular form of debt servitude that is required of the nation's women: "It is in the nuclear family that the most crucial process of defiance and faith is centered. . . . Here emerge the most indispensable acts of capital formation: the psychology of giving, saving and sacrifice, on behalf of an unknown future, embodied in a specific child—a balky bundle of possibilities that will yield its social reward even further into time than the most foresighted business plan" (Gilder 1986, 198–99).

It is no accident then that the counteractive tendencies of neoliberal conservatism come to a head on the question of embryonic life and its scientific regeneration. The stem cell line seems to offer up the most radical materialization of the evangelical faith and its promise of an endlessly renewable surplus of life. At the same time it threatens to undermine the very precepts of normative reproduction and therefore needs to be recaptured within the social and legislative limits of the potential person—and its right to life.

THE UNBORN BORN AGAIN:
THE RIGHT-TO-LIFE AND BORN-AGAIN MOVEMENTS

The movement that we now recognize as born-again evangelical Christianity underwent an extraordinary reawakening in the early 1970s. In its revived form, the evangelical movement took up the Protestant ethic of self-transformation—impelling its believers to be born again, in a kind of personal reenactment of Christ's death, burial, and resurrection—and turned it into something quite different in scope. What distinguished this movement from both mainline Protestantism and earlier evangelical revivals was its intense focus on the arena of sexual politics and family values. Faced with a rising tide of new left political demands, from feminism to gay rights, the evangelical movement of the 1970s gave voice to a newfound nostalgia—one that obsessed over the perceived decline of the heterosexual, male-headed, reproductive white family. The concerns of the right-to-life movement have ranged from the opposition to equal opportunities and domestic violence legislation to gay marriage. But if there was one issue that focalized the energies of the early movement it was the *Roe v. Wade* decision of 1973, in which the U.S. Supreme Court voted to overturn state bans on abortion. As one editorial of the late 1970s pointed out, *Roe v. Wade* was the "moment life began conception—'quickening,' viability, birth: choose your own metaphor—or the right to life movement" ("The Unborn and the Born Again," 1977, 5). The born-again evangelical right was reborn as a crusade to save the unborn.[13]

We now so commonly associate the evangelical right with a "pro-life" politics that it is difficult to recognize the novelty of this revival. Their obsessive focus on the question of abortion was, however, unprecedented within the history of Protestant evangelicalism—so much so that the early neo-evangelicals borrowed their pro-life rhetoric from orthodox Catholicism, if only to later rechannel it through distinctly mass-mediated, populist, and decentralized forms of protest.[14] In the process the evangelical right brought a new element into its own traditions of millenarianism and born-againism. For evangelicals awaiting the millennium, the unborn came to be identified with the last man and the last generation—indeed the end of the human race. At the same time it was this last—and future—generation that most urgently required the experience of conversion or rebirth. The evangelical tradition had long identified the unsaved soul with Christ before the resurrection, but now both were being likened to the unborn child in utero. In the born-again how-to tracts of the 1970s, Christ himself had become the unborn son of God, while we were all— prior to salvation—the fetal inheritors of the Lord.[15] In this context of tortuous temporal amalgamations it was no surprise that the question—can the unborn be born again?—emerged as a matter worthy of serious doctrinal debate.

At the outset the evangelicals understood the pro-life movement to be a project of national restoration. The United States was founded on religious principles— indeed on the principle of the right to life—according to the new evangelical right. *Roe v. Wade*—a decision that after all was most likely to affect young white women—was decried as an act of war that threatened to undermine the future reproduction of the (white) American nation, its possibility of a redemptive afterlife.[16] It was also the last and fatal blow in the protracted process of secularization and pluralism that had led to the decline of America's founding ideals. *Roe v. Wade* had emptied the gift of life of all foundation—the future existence of America had been effectively undermined, offered up in a precarious, promissory form, a promise that might never be redeemed. Ontologically, it seemed, America was suspended in the strange place that is also reserved for the frozen embryo (hence an obsessive focus not simply on the unborn but more particularly on the frozen or in vitro unborn).

At the same time, and characteristically for the evangelical right, these concerns about the sexual and racial reproduction of the American nation came together with a sense of malaise in the face of America's growing state of indebtedness. As the evangelical Pat Robertson (1991, 118) has remarked: "Any nation that gives control of its money creation and regulation to any authority outside itself has effectively turned over control of its own future to that body." Here the idea that the reproducers of the unborn nation might be at risk of defaulting feeds into the fear that the United States' economic future might be similarly imperiled, suspended as it were on the verge of a promise without collateral. Thus, along with its enthusiastic support for U.S. debt imperialism, the evangelical right also gives voice to the suspicion that the economic reproduction of the United States is becoming dangerously precarious, promissory, contingent, a matter of faith—in urgent need of propping up.[17] The nightmare of someone like Pat Robertson is that the promissory future of U.S. debt may not be restored within the territorial limits of America itself, that the future may fail to materialize within the proper limits of self-present nationhood, here and now. And because he understands that the nation lies at the nexus of sexual and economic reproduction, he calls for a politics of restoration on both fronts.

Delirious as it may seem, the religious right at least recognizes that from the point of view of traditional state financing, the postmodern American nation is literally poised on the verge of birth—unborn—its future contingent on the realization of a debt that has not yet and may never come to maturity. Their fear is that its potential may be realized in the form of excess, escaping appropriation. And in anticipation of this threat, they call for a proper rebirthing of the unborn, the resurrection of a new man and a new nation, from out of the future. But what would it mean to refound the future? In what sense is it possible to rebirth the unborn? It is in the form of this temporal ellipsis that the right-to-life movement articulates its politics of nationhood: what needs to be restored is of course the foundational moment of America, the act through which the Founding Fathers inaugurated the nation, but this moment is itself constitutive of the right to life of the unborn, contingent, in other words, on the return of the "not yet."

The pro-life movement has invented an extraordinary number of ritualistic methods for memorializing this contingent future: from online monuments to the unborn to court cases undertaken on behalf of the future victims of genocidal abortion. Herein lies the novelty of (neo-)fundamentalism, of fundamentalism for the neoliberal era: in the face of a politics that operates in the speculative mode, fundamentalism becomes the struggle to reimpose the property form in and over the uncertain future. This property form, as the right-to-life movement makes clear, is inextricably economic and sexual, productive and reproductive. It is ultimately a claim over the bodies of women. Except here the name of the dead father is replaced by the image of the unborn child as sign and guarantor of women's essential indebtedness.

Under Reagan the rhetoric of the pro-life movement, with its rewriting of the Declaration of Independence as a right-to-life tract, entered into the mainstream of American political discourse, so that a hard-line conservative such as Lewis E. Lehrman (1986) could declare that the moral and political restoration of America would depend on the Republican Party welcoming the unborn "in life and law." Reagan himself, however, failed to live up to the expectations of his moral electorate, and it was not until George W. Bush came to power that the pro-life movement acceded to anything like a real presence within the decision-making processes of government. When it did so, it was after making a detour via the neoconservative right. Throughout the 1990s, a period when both moralist and militant extremes of conservative thinking were on the backfoot, a second generation of neoconservatives began to make overtures to the religious right, inviting pro-life representatives to work at their think tanks while they themselves began to issue very public declarations linking the political and strategic future of the American nation to its upholding the "founding" principle of the right to life.[18]

Since then, pro-lifers and neoconservatives have joined forces in mounting a more general assault on all kinds of embryo research, particularly in the area of stem cell science. It was no surprise when a neoconservative Catholic such as Michael Novak (2001) came out in opposition to Bush's compromise stem cell decision of 2001, to announce that it threatened the unborn potential of America, and by extension the future salvation of the rest of the world: "This nation began its embryonic existence by declaring that it held to a fundamental truth about a right to life endowed in us by our Creator. The whole world depends on us upholding that principle." But the 1990s had also seen more mainline, previously "secular" neocons such as William Kristol launching himself into the arena of right-to-life politics, in a series of impassioned stay-of-execution pleas on behalf of the unborn. For Kristol the connection between a muscular, neo-imperialist foreign policy and a pro-life position was clear: what was at stake in both cases was the restoration of an emasculated America, the rebirth of its unborn nationhood. As Kristol and his coauthor George Weigel (1994, 57) have written: "We will work to build a consensus in favor of legal protection for the unborn, even as we work to build an America more hospitable to children and more protective of families. In doing so, our country can achieve a commitment to justice and a new birth of freedom."

It is probably too early to assess the long-term consequences of these developments, but at the very least it might be ventured that the alliance between the neoconservatives and the Christian right has brought a new and alarmingly literal legitimacy to the warmongering, millenarian, and crusading rhetoric of the right-to-life movement. After all, as commentators of the Christian right have detailed, pro-life representatives are now occupying key advisory positions at every level of U.S. government, including foreign policy, and now dominate U.S. delegations to the UN, where they frequently form voting blocs with the Vatican and the strictest of Islamic countries.[19] The most obvious effect of this presence so far has been in the arena of foreign aid, where U.S. federal funds are now indexed to stringent antiabortion, antiprostitution, anticontraception, and abstinence guidelines.

On a rhetorical level, too, George Bush has consistently drawn together the language of the Christian right—with its evocations of a war on the unborn, and its monuments and memorials to the unborn—with the newly legitimized, neoconservative defense of just war. Is this the harbinger of a new kind of war doctrine? One that returns to the doctrine of just war theory, while declaring justice to be without end? One that speaks in the name of life, like humanitarian warfare, while substituting the rights of the unborn for those of the born?

Certainly, this has been the subtext of Bush's official declarations on America and the culture of life.[20]

As a counter to these slippages, it is important to remember that the most immediate precedent to the terrorist attacks of September 11, 2001, can be found in the string of bombings and murders committed by homegrown right-to-life groups and white supremacist sympathizers over the past few decades.[21] These attacks have attracted nothing like the full-spectrum military response occasioned by September 11. On the contrary, one of the ironies of Bush's war on terror is that it is being used as a pretext for bringing the culture of life to the rest of the world. In this way, even as it emanates from the precarious center of debt imperialism, Bush's politics of life works in tandem with the many other neofundamentalist movements of the neoliberal era.

EPILOGUE

THE FIRST CHAPTER OF THIS BOOK BEGAN IN THE ATMOSPHERE OF INTENSE
paranoia and speculation that accompanied the American economic crisis
of the 1970s. At this time the object of its paranoia was Japan, whose newly
energized economy was credited with having invented the methods of post-
Fordism and flexible accumulation. And although most analyses of the time
would have singled out electronics and digital technologies as the key industries at stake in the competitive stand-off between North America and its new
rivals, I have argued that one of the principal responses to this crisis was the
reorganization of U.S. life science production along commercial and highly speculative lines. As I complete this book, it seems to me that the United States is
facing a new geography of imperialist power relations and threats (imagined
or otherwise)—this time from the likes of China and India rather than Japan.
With their growing interest in the biosciences and massive industrial *and*
postindustrial capabilities, these rivals threaten to materialize the still speculative possibilities of the North American biotech enterprise.

As an insight into these possibilities, a number of key developments can be
singled out. First, North American– and European-based pharmaceutical and
biomedical enterprises are increasingly looking to outsource and offshore their
clinical trials to India and China, where (reading between the lines) the ethical costs of human life come cheaper. This trend toward the offshoring of biomedical and clinical labor, along with the emergence of transnational markets
in "donated" organs, blood, tissues, and eggs, points to the new divisions of
labor, life, and surplus that are likely to accrue around a fully fledged bioeconomy. However, the trend is not limited to immediately corporeal forms of
hard "service" labor, such as those required by participation in clinical trials,
but is also emerging as a possibility in the area of scientific knowledge production and lab work. With its workforce of highly educated (and again,
underpaid) scientists, it is feasible that in the not too distant future, even the

labor of North American–based "symbolic analysts" will be relocated offshore. Over the past few decades China itself (following the example of India) has massively increased its investment in all areas of life science production—from agricultural to biomedical technologies—and is beginning to assert itself not merely as a supplier of labor and tissues but as a technological leader in its own right. This is a development that at the very least challenges the postindustrial utopias of such early neoliberals as Daniel Bell (for whom the service and knowledge economies were to be the last refuge of privileged U.S. workers) and, as such, promises to rework the dynamics I have outlined in this book.

The shifting dynamics of world life science production raise a number of questions that I can only anticipate here. What are the distinctive forms of Chinese "neoliberalism" and what is its relationship to imperialism?[1] Moreover, what are the cultures of life, health and medicine, informing China's investment in the field of life science production? The recent history of Chinese biopolitics, even in its neoliberal forms, is strikingly different from that of North America and Europe and thus promises to unsettle the current status quo of international political and ethical relations.[2] But even this configuration of powers is being refashioned in response to new forms of biomedical labor, new demands (the "ethical" imperatives of foreign investors and regulatory agencies), new desires, and undoubtedly new modes of contestation.

It is possible, in other words, that the still emergent and highly speculative markets in life science technologies will end up consolidating in forms that were not foreseen by the United States and will be as responsive (if not more) to developments in China, India, and other emerging economies as to the internal politics of the United States. If this is the case, a very different politics of life, labor, and resistance is sure to emerge from the vestiges of the biotech utopia.

Notes

◻ *INTRODUCTION*

1. Throughout this book I use the terms "liberal" and "neoliberal" in the specific sense of the classical liberal and the neoliberal political economies. I therefore wish to distinguish my use of the term "liberalism" from the more colloquial American understanding of liberalism, as well as the philosophical currents associated with it as a moral discourse. Although it is true that the term "neoliberalism" was taken up much earlier in the non–English-speaking Western European countries, it is now becoming familiar to English speakers too.

2. Italics mine.

3. This is not to say that the concept of reproduction is ever fully analyzed by Marx and Foucault. In this regard philosopher of science Ludmilla Jordanova's 1995 work serves as an illuminating corrective. Taking up Foucault's hypothesis of the late-eighteenth-century invention of life, she notes that this period also saw the first recognizably modern formulation of the concept of "reproduction."

4. All translations from the original French are my own.

5. This is not to say that the question of welfare state biopolitics has been exhaustively explored by Foucault and F. Ewald. On the contrary, it seems to me that this work unduly neglects the role of social and biological reproduction within the calculations of Keynesian and Fordist growth strategies. What has come to be known as the "family wage" is only another way of saying that the Keynesian social state more or less forcefully relegated a whole sector of the female population (if only the middle class and in the developed world) to the role of state-supported reproductive labor. This "gift" of female, reproductive labor—which second-wave feminist theorists have so forcefully deconstructed—is really the focal point of the welfare state's "economy of life." Its role is analogous to that of the reserve bank, in that it furnishes a reserve of (biological) wealth that needs to be maintained outside of the sphere of productive labor and commodity exchange, while nevertheless functioning as the necessary condition and determining value of all exchange relations. Moreover, as postcolonial theorist Laura Stoler (1995), among others, has argued, Foucault's analysis of state biopolitics pays hardly any attention to the practices of colonialism, let alone the emerging power relations of the postcolonial era. Foucault's few developments on

race and eugenics are largely concerned with the Nazi eugenic state and offer little insight into the resonances between European state racism and the history of imperialism. Finally, any detailed exploration of nation-state biopolitics as such would need to go further than Foucault and Ewald in exploring the biomedical and reproductive sciences developed in the course of the twentieth century. As coauthors Catherine Waldby and Robert Mitchell have argued in *Tissue Economies* (2006), Richard Titmuss's classic 1971 work on blood donation, which establishes the public blood bank and free redistribution of blood as the founding principle of welfare state nationalism, is a key text in this regard. Again, it could be argued that Titmuss's work reveals the foundational role of the human biological within the growth strategies of the Keynesian nation-state. Titmuss himself prefers to use the language of the gift—but a less idealistic reading would conclude that the welfare state establishes the donation and redistribution of blood as a national biological reserve, in much the same way as it requires the unpaid reproductive labor of women. The public, nation-based blood bank is the biological equivalent of the national reserve bank. It is only "outside of exchange" to the extent that it constitutes the fundamental value underlying all exchange relations.

6. On this point I am in accord with the following statement from Foucault's *Naissance de la biopolitique*: "In American neo-liberalism, the point is . . . to generalize the economic form of the market. The point is to generalize it across the entire social body and even the entire social system which, normally, isn't included in or sanctioned by monetary exchange. This absolute generalization, this unlimited generalization of the market form, as it were, entails a certain number of consequences. . . . Firstly, the generalization of the economic form of the market beyond the sphere of monetary exchange, functions, in American neo-liberalism, as a principle of intelligibility, a principle for decoding social relations and individual behavior. What this means is that a market-based analysis, an analysis in terms of supply and demand, comes to serve as a schema that can be applied to non-economic domains. And thanks to this analytic schema and grid of intelligibility, it will become possible to reveal a certain number of intelligible relations at work in non-economic processes and relations, which wouldn't otherwise be apparent—a sort of economistic analysis of the non-economic" (2004, 248–49; my translation).

7. The French sociologist Jean Gadrey (2003) has argued that the concept of economic growth only comes into its own with the advent of Keynesianism and the methods of mass, Fordist production. Moreover, historians of development theory—with its language of first, second, and third world—have similarly argued that the imperialist politics of growth finds it fullest expression in mid-twentieth-century international relations discourse. See on this point Rist 2004.

8. In this respect Randy Martin's 2002 study *Financialization of Daily Life*, with its analysis of the convergence between postmodern psychologies and speculative strategies of risk management, comes much closer to the kind of critique I am attempting here.

1 ◻ *LIFE BEYOND THE LIMITS*

Epigraph: Rothschild and Mancinelli 2001, 1092.

1. For illuminating readings of this period and the genre of futures analysis, see Ross 1991, 169–92; and Buell 2003, 177–246.

2. Famously, philosophers Gilles Deleuze and Félix Guattari take delirium as the starting point for rethinking both capitalism and desire. See on this point the interviews with Deleuze and Guattari in Guattari 1995, 53–92. But see also the 1995 work of a more orthodox Marxist theorist, Daniel Bensaïd, *La discordance des temps: Essais sur les crises, les classes, l'histoire,* for a closer textual analysis of the concept of delirium in Marx's work. Bensaïd suggests that for Marx the problematic of delirium is intimately connected with capitalism's moments of crisis and transformation. While most contemporary readings of Marx focus on the first volume of *Capital* ([1867] 1990) and the analysis of commodity fetishism, the *Grundrisse* ([1857] 1993) and the third volume of *Capital* ([1894] 1981) are more directly concerned with debt creation and crises of overproduction. I choose to read Marx from the point of view of delirium, rather than fetishism, because it moves us away from a representational theory of time and money. The question is no longer whether the sign (money) adequately represents the use-value of labor (as fundamental value), but rather under what conditions are the world-transformative possibilities of collective labor separated from their power to act. In other words the concept of delirium moves us closer to a creative philosophy of time, in which time becomes no more or less than the immanent transformative force of matter. In this respect my reading of Marx is closely aligned with that of Italian autonomist thinkers such as Antonio Negri. I explore the implications of Negri's approach to Marx in my article "Marx Beyond Marx" (Cooper 2007).

3. See in particular Giovanni Arrighi's account (2003, 62–69).

4. See Kenney 1986, 191–93; and Drahos and Braithwaite 2002, 154–55.

5. See Buttel, Kenney, and Kloppenburg 1985.

6. See Pignarre 2003, 26–62; and Drahos and Braithwaite 2002.

7. See Kenney 1986, 197–98.

8. See Buttel, Kenney, and Kloppenburg 1985, 39.

9. An overview of mergers and acquisitions that had taken place at the time of completing this manuscript can be found in Razvi and Burbaum 2006.

10. These funds may operate alone or as part of a larger corporation and specialize in financing the late stages of a business enterprise, right up until the moment when it goes public at an IPO or when it is sold to a larger firm through a trade sale. Their investment decisions are indexed to public mood rather than to the fundamentals of a given business, since they tend to invest in companies that have little or no tangible capital, withdrawing from the venture as soon as it is brought to market. Venture capital flourishes in moments of high public faith in the promises of science. The more the public believes in the promise of a given enterprise, the higher the valuation it will make at an IPO.

11. On this point see Jessop 2002, 100–1.

12. On the counterproductivity of capital, see Marx [1857] 1993, 414–23.

13. This point is made most forcefully in Loeppky 2005.

14. See Dickson 1984 for an overview of this crisis discourse and its effects on U.S. science policy.

15. On all these points, see Zeller 2005. For details on the redistribution of life science funds, see Estes and associates 2001, 51–93.

16. On all these points see Coriat and Orsi 2002; and Zeller 2005.

17. See Fortun 2001 for a fascinating analysis of the life science promise.

18. This is in part due to the prior absence of any clear distinction between invention and discovery in American common law tradition. But it can also be attributed to the U.S. government's efforts, from Reagan on, to actively promote intellectual property rights (IPR) as a key element in the United States' economic future. To this end, a special appeals court for patents and trademarks was set up in 1982. This institution was largely responsible for the flood of biological patents approvals over the following decades. On these points, see Coriat and Orsi 2002.

19. On all these points, see Chesnais and Serfati 2000.

20. Whether this particular form of imperialism will be able to sustain itself for any length of time, and at what price, is a matter of considerable debate, and I make no attempt to offer any forecasts here. However, there does seem to be a general consensus that the synergy between export-surplus countries and the U.S. budget and trade deficit has become structural to world economic relations. In other words any change to the status quo would have world-systemic consequences.

21. On the capitalist delirium, see Marx [1894] 1981, 466, 470, 515–17. On national debt Marx offers the following comment: "In the way that even an accumulation of debts can appear as an accumulation of capital, we see the distortion involved in the credit system reach its culmination" (ibid., 607–8).

22. Here I am extrapolating from Marx's comments on the world-expansive tendencies of capital and its specific space-time. In the *Grundrisse*, for example, Marx writes that "the tendency to create the *world market* is directly given in the concept of capital itself" ([1857] 1993, 408). There is no systematic Marxian theory of world order. However, it seems clear that if there is a world system for Marx, it differs markedly from the more familiar concepts of world image, metric imperial space, and global linear space offered by Martin Heidegger ([1938] 1977), Rosa Luxemburg (1973), and Carl Schmitt ([1950] 2003), respectively. In the briefest terms the Marxian understanding of capital as debt, taken to its extreme conclusions, requires a nonmetric and nonrepresentative conception of space-time. The temporality of debt is creative—auto-productive—rather than representative. Moreover, as the *Grundrisse* makes clear, the capitalist debt form tends to deflect from all limits and mediations. Its mode of self-differentiation is fractal rather than dialectic. It therefore requires another philosophy of time and matter. There are inklings of this alternative philoso-

phy in Marx's work—evident in particular in his abiding interest in Lucretian and Epicurian materialism. It seems to me that a debt-inspired philosophy of time-matter comes to fruition in the work of Henri Bergson, Gilles Deleuze, and such philosophers of science as Ilya Prigogine and Isabelle Stengers—making their work both irresistible and difficult (not impossible) to use in terms other than descriptive. A more recent discussion of Marx's creative conception of world-time can be found in Jean-Luc Nancy's fascinating 2002 study. He notes that "the world has now escaped from representation, from its own representation and from a world of representations, and it is no doubt here that we reach the most contemporary determination of the world. Already with Marx, the world as deployment of the self-production of man prescribed this escape from representation (even though 'production' undoubtedly still bears traits of representation in his work)" (Nancy 2002, 38, my translation from the original French). It seems to me, however, that the problematic of self-production or auto-valorization in Marx's work requires a fuller investigation of debt and its powers of violence. Failing this, Nancy ends up offering something like a description of contemporary debt production as if it were liberating in itself. His attempt to distance himself from Marx thus ends up hinging on the distinction between the (Hegelian) notion of the bad infinite of accumulation and the collective production of "the inaccumulable or the inequivalent" (ibid., 43). Despite Nancy's protestations, it is precisely the latter understanding of production that corresponds most closely to Marx's thinking on surplus value. Marx's mathematics are not Hegelian. The question of resistance thereby becomes considerably more complex.

23. For a discussion of the difference between recombinant DNA and traditional breeding techniques, see Sapp 2003, 234–51.

24. In this way rDNA makes use of a biological process—infection—which nineteenth-century biology considered solely from the point of view of the pathological. Writing in the early 1960s, the microbiologist René Dubos anticipated the importance of these developments, predicting that the germ theory of disease would at some point be complemented by a theory of *creative infection*. The new biosciences, in other words, would effect a fundamental redefinition of the boundaries between the normal and the pathological, the sterile and the productive. On this point, see Dubos 1961.

25. See Le Méhauté 1990 for a useful introduction to the mathematics of fractals and its relationship to the physics of dissipative structures. A fractal is a curve that is not "rectifiable"—in other words a curve that tends toward no finite limit. In more imagistic terms we might say that the fractal is a curve that continuously produces discontinuity. Such curves were described as pathological by nineteenth-century mathematics. They were formalized in the twentieth century by the French mathematician Benoit Mandelbrot.

26. For a complete history of NASA's exobiology and astrobiology programs, see Dick and Strick 2004. On Lovelock's role in the program and the invention of the Gaia hypothesis, see ibid., 82–84. On the recent restructuring of the program

that I am interested in here, see especially ibid., 202–20. For an overview of the theoretical perspectives currently being developed within the NASA astrobiology program, see the 2003 work of microbiologist Charles S. Cockell, *Impossible Extinction*.

27. For an overview of evolutionary economics and the return to evolutionary models in economic theory, see the 2005 collection edited by Kurt Dopfer, *The Evolutionary Foundations of Economics*.

28. For an in-depth analysis of the conference and its influence on economic theory, see Mirowski 1996. See also Helmreich 2001 on the relationship between Santa Fe artificial life theory and new economy business models.

29. For a detailed account of such innovations, and related proposals, see Daily and Ellison 2002 as well as Chichilnisky and Heal 1998.

30. For a detailed response to the Energy Policy Act, see Goozner 2006.

31. On this point, see Pirages and Cousins 2005.

2 ❐ ON PHARMACEUTICAL EMPIRE

Epigraph: Natsios 1997, 1.

1. For a detailed critical response to the securitization of AIDS, see Elbe 2005.

2. On the definition of "complex emergencies" and their implications for U.S. foreign policy, see Natsios 1997.

3. See WTO 1996, available online at
http://www.wto.org/english/tratop_e/trips_e/t_agm3_e.htm.

4. On this point, see Resnick 2002.

5. Compulsory licensure allows a country to license the domestic production of a product without the patent holder's permission. Parallel importing allows a country to import pharmaceuticals from the cheapest foreign market without having to negotiate a direct contract with the original producer.

6. On the inside story behind the TRIPs agreement, see Sell 2003 as well as Drahos and Braithwaite 2002.

7. On all these points, see Sell 2003 as well as Drahos and Braithwaite 2002.

8. For an insight into the creation of the "intellectual property crime," see Fraumann 1997.

9. On these points, see Peberdy and Dinat 2005 as well as Williams et al. 2002.

10. For classic studies on the connections between theories of immunity and political sovereignty, see Waldby 1996 and Esposito 2002.

11. For a preliminary insight into some of these shifts in conceptions of immunity, see Varela and Coutinho 1991.

12. On this point, see Rothschild 1995.

13. On the securitization of migration, see Didier Bigo's classic article "Security and Immigration" (2002). On the connections between AIDS and migration policy, see Haour-Knipe and Rector 1996.

14. See Claudia Aradau's fascinating article on this topic, "The Perverse Politics of Four-Letter Words" (2004), where she looks at the paradoxical position of migrant sex workers at the meeting point of humanitarian and security discourses.

15. On this point, see Youde 2005.

16. The social critic Susan Sontag (1988, 62) quotes the South African foreign minister Botha as warning that the "the terrorists are now coming to us with a weapon more terrible than Marxism: AIDS."

17. Again, see Youde 2005, 426–27.

18. South African Department of Defense, *South African Defense Review*, chapter 1, introduction, available online at http://www.dod.mil.za/documents/defencereview/defence%20review1998.pdf.

19. On this point, see Ashforth 2005, 104–5.

20. See in particular ibid. Also see Comaroff and Comaroff 1993.

21. On this point, see Comaroff and Comaroff 2001.

22. For this reason I do not share anthropologist James Ferguson's optimism with respect to the political potential of popular moralisms. See on this point Ferguson 2006, 69–88. On the contrary, such counterimperialisms have a tendency to reinstate private property at the level of sexual, familial, and moral relations.

23. On xenophobia and rising levels of violence against migrants in South Africa, see Crush and Pendleton 2004.

24. On these points, see Arnott 2004 as well as Slaughter 1999.

25. After going on an antiabortion and proabstinence crusade, Bush's Global AIDS Fund now obliges foreign NGOs receiving U.S. HIV and antitrafficking funds to sign a pledge opposing prostitution. For a detailed account of Bush's AIDS and global public health policy, see Kaplan 2004, 167–93 and 219–43. On the more recent antiprostitution pledge, see Schleifer 2005.

3 ☐ PREEMPTING EMERGENCE

Epigraph: Mandelbrot 2004, 41.

1. These processes of "horizontal gene transfer" include transduction (viral infection between bacteria), transformation (the direct uptake of a DNA sequence from the environment), and conjugation (involving cell-to-cell contact and mobile pieces of extrachromosomal DNA called plasmids). Research into horizontal gene transfer boomed in the late 1980s and 1990s. For one of the first overviews, see Levy and Novick 1986. See also Miller and Day 2004.

2. Certain biologists argue that the sudden upsurge in microbial resistance from the mid-1970s on cannot be ascribed to the overuse of antibiotics alone, and may well be linked to the commercial-scale use of transgenic organisms. See on this point Ho 1999, 181–82, 192–200.

3. There exists an extensive body of international relations theory arguing for the concepts of human, biological, and microbiological security (as well as other varia-

tions such as food and water security). See in particular Chyba 1998, 2000, and 2002. See also Brower and Chalk 2003.

4. Quoted in Lefters, Brink, and Takafuji 1993, 272. See also Davis 2005, 129.

5. For an overview of the concept of the catastrophe risk in these three domains, see F. Ewald 1993 and 2002, Bougen 2003, and Haller 2002, respectively.

6. See Natsios 1997, 2–6.

7. Italics mine.

8. This is not to suggest that the BTWC was ever successfully enforced. Biowarfare expert Susan Wright (1990) has pointed out that from the beginning the BTWC lacked an enforcement protocol and allowed room for research and limited stockpiling. Under Reagan there had been a return to bioweapons research in the United States, although this was considerably stepped up under Clinton. Ironically, although the Nixon administration was responsible for withdrawing the United States from bioweapons research, it was Nixon who launched the first war on drugs, initiating a campaign of transnational counterinsurgency whose modus operandi in many ways anticipates today's war on terror. See McCoy 2003, 387–460.

9. The most pertinent references here are Carter 2002 as well as Carter and White 2001 because of their interest in the concept of "catastrophic terrorism."

10. The concept of "catastrophic terrorism" was promoted in the late 1990s by Clinton's assistant secretary of defense, Ashton Carter, among others, and has become a commonplace of U.S. defense discourse since the attacks of September 11, 2001.

11. According to the Pentagon's draft "Defense Planning Guidance" for 1994 through 1999, drawn up in 1992, the first objective of the United States in the post–Cold War era should be to "prevent the re-emergence of a new rival, either on the territory of the former Soviet Union or elsewhere, that poses a threat on the order of that formerly posed by the Soviet Union." Quoted in "Excerpts from Pentagon's Plan" 1992.

12. In *Multitude: War and Democracy in the Age of Empire*, coauthors Antonio Negri and Michael Hardt (2004) suggest that an alternative genealogy of current U.S. strategy can be traced back to the Nixon era and the beginnings of the "neoliberal revolution." In this sense the Nixon administration can be situated at the crossroads between two eras of warfare. While on the one hand Nixon continued to aggressively uphold the Cold War status quo against the emergence of newer kinds of enmity, the United States was already engaging on the sidelines in its own politics of counterinsurgency, from Vietnam to Latin America to the war on drugs (surely an early form of biowarfare) (ibid., 38–40). It is these counterinsurgent tactics that have now come to dominate U.S. defense strategy.

13. On this point see Müller and Reiss 1995, 139–50, which notes that many within the Clinton administration "feared that the United States, as the world's lone superpower, was now devising the means to unilaterally and preemptively destroy the nuclear programs of countries in the developing world." The authors go on to

note that "although some Pentagon officials privately admit that counter-proliferation still envisions preemptive military strikes, more senior officials, especially Assistant Secretary of Defense Ashton Carter, have explicitly and repeatedly disavowed any such role" (ibid., 139). The difference between U.S. defense under Clinton and that under the influence of the neoconservatives can be pinpointed in the latter's willingness to unequivocally embrace preemption.

14. Conversely, the "emergent" nature of the terrorist threat has been used to justify the United States' relative inaction prior to the attacks of September 11. "When was 9/11 imminent?" was the rhetorical question put forward by George Bush's neoconservative faction. For an extended commentary on this far-reaching shift in the understanding of preemption, see O'Hanlon, Rice, and Steinberg 2002. It should be noted that the neoconservative understanding of preemption is indebted to the work of military strategists Albert and Roberta Wohlstetter, particularly the latter's study *Pearl Harbor: Warning and Decision* (1962). This work is concerned with the limits of mutual deterrence when faced with situations of unpredictable surprise and represents a very early argument in favor of the doctrine of preemption. Significantly, Roberta Wohlstetter's work includes an extensive discussion of the psychology of future-oriented perception, surprise, and the operative power of "wishfulness."

15. Italics mine.

16. Italics mine.

17. The report was written by Peter Schwartz, CIA consultant and former head of planning at Royal Dutch/Shell, and Doug Randal of the United States–based Global Business Network. For further details on the report, see Townsend and Harris 2004.

18. For details of this and other programs in the biological sciences, see the report "Biological Sciences" of the DARPA Defense Sciences Office, available online at http://www.darpa.mil/dso/thrust/biosci/biosci.htm (accessed March 2006). See also John Travis's 2003 "Interview with Michael Goldblatt, Director, Defense Sciences Office, DARPA." In this interview Goldblatt notes that the "original focus of the DARPA efforts in biological warfare defense were aimed at protection from genetically engineered threats—where you have to protect against the unknown and perhaps unknowable" (Travis 2003, 158).

19. DARPA is not alone in its preemptive vision of biowarfare. In a recent article overviewing the current state of bioweapons research in the United States, Susan Wright (2004, 60) points to a general trend toward "pre-emptive" visions of biodefense, where the aim is "to defend not only against *known* pathogens but also against *futuristic* ones—genetically altered microbes that could overcome existing vaccines or antibiotics or attack the immune system in novel ways, and so forth."

20. Two excellent studies explore the flourishing role of the private sector in security operations, including humanitarian interventions: Avant 2005 and Singer 2003.

21. For an unrivaled analysis of Hurricane Katrina as an episode in "punctuated social evolution," see Caffentzis 2006.

22. On the history and relevance of Posse Comitatus, see Healy 2003, a report sponsored by the right-wing libertarian Cato Institute. On Bush's response to bird flu, see CNN 2005 and Healy 2005.

23. My translation from the original Italian.

24. Not surprisingly, some of the best accounts of the so-called new economy are retrospective ones. See, for example, Henwood 2003.

25. First announced in 2002, Bush's BioShield Project was stalled in Congress for over a year and received a less than enthusiastic response from the pharmaceutical and biotech companies it was supposed to entice. The final version of the project not only contained generous funding provisions for the creation of medical responses to bioterrorist attack, but also measures allowing for the fast-tracking of clinical trials and federal drug approval. For full details of the BioShield Project, see White House 2003. For a more extensive account of U.S. legislation on bioterrorism from the closing years of the Clinton administration on, see Wright 2004; and on the Bush years, see Guillemin 2004.

26. On the differences and continuities between Clinton's "new economy" and the era of permanent warfare, inaugurated with September 11, see in particular Marazzi 2002 as well as Mampaey and Serfati 2004. Coauthors Luc Mampaey and Claude Serfati (2004, 250) note that "after the wars in Afghanistan and Iraq, American 'markets' are perhaps beginning to 'internalize' into their behavior the inevitability of new wars and military operations; to forge as it were a convention based on the idea of 'war without limits,' in which the *discretionary* use of military force by the US represents their new horizon." More forcefully, Marazzi (2002, 154) argues that "the war . . . against terrorism represents *the continuation of the New Economy by other means*" (italics mine).

27. It should be noted that Clinton was already moving in this direction. In the late 1990s the Clinton administration introduced new counterterrorism laws (blurring the difference between military emergency and domestic law enforcement), while approving a massive increase in counterterrorism funds (a sizable portion going to bioweapons research). On this point, see Hammond 2001–2 as well as Miller, Engelberg, and Broad 2001, 287–314. According to science journalist Edward Hammond (2001–2, 42), when Clinton turned toward biodefense research in the late 1990s, it was not only in response to the Pentagon but more important to lobbying from the ailing genomics sector, which was looking for an emerging market to invest in as the genome sequencing projects came to an end.

28. Much could be said about the Bush administration's efforts to subcontract social welfare, including emergency response, to faith-based initiatives. In this area the Republican right seems to have learned from neofundamentalist Islam that the postcatastrophic landscape is the perfect breeding ground for reaction. It is interesting that Pat Robertson's Operation Blessing was one of the principal faith-based charities to be listed on the official FEMA Web site in the wake of Katrina.

4 ◻ CONTORTIONS

Epigraph: Simondon 1995, 222–23.

1. On this point, see Stocum 1998, 413–14, which argues that "the aim is to . . . recreate an embryonic (regenerative) environment in an injured adult tissue."

2. "An organ is essentially an enduring thing. It is a movement, a ceaseless change *within the frame of an identity*" (Carrel and Lindbergh, 1938, 3, italics mine).

3. On dissection, vivisection, and organ transplantation, see ibid., 1–6, 219—21.

4. On the history of organ preservation techniques, see Rubinsky 2002, 27–49, as well as M. Phillips 1991.

5. My translation from the original French.

6. On these three methods, see Auger and Germain 2004.

7. My translation from the original French.

8. The architectural method of TE is thus in keeping with philosopher Brian Massumi's description of the "biogram" as abstract generative condition of bodily experience. "The relational, variational continuum pertains to a qualitative space that can only be described topologically. Its recursivity cannot be ignored, so it is as immediately a non-linear temporality as it is a non-Euclidean space" (Massumi 2002, 197).

9. On this point, see M. Phillips 1991.

10. See "Is the Product a Medical Device?" available online from the U.S. Food and Drug Administration at http://www.fda.gov/cdrh/devadvice/312.html (accessed March 2006).

11. There is a real overlap between computer-assisted design in architecture and the emerging field of computer-aided tissue engineering (CATE). See Sun, Darling, Starly, and Nam 2004; as well as Sun, Starly, Darling, and Gomez 2004.

12. See, for example, the work of D'Inverno, Theise, and Prophet 2005.

13. How could we characterize the minimal conditions of nonmetric time? Following Klein's formulations for topological space, we would have to imagine a temporality in which each "instant" or "present" is continuous with all others, so that the past continuously morphs into the future "at infinite speed," while at the same time escaping all immediate presence. A perpetual continuity of the past and the future, of the unborn and the born again; embryogenesis as process without progression. This is an understanding of time that comes very close to the Deleuzian concept of becoming. In *Logic of Sense*, for example, Deleuze (1990, 80) describes the becoming of the event as that which is "always forthcoming and already past." Simondon (1995, 223) also calls for a philosophy of topological time, arguing that a "true" biotechnology would be one that strives to regenerate the body in nonmetric space and nonchronological time. This vision of the future possibilities of biomedicine can be usefully read alongside Simondon's philosophical characterization of life as a process of perpetual birthing: "The individual concentrates within itself the dynamic which led to its birth and perpetuates this first operation in the manner of a continuous individuation; to live is to perpetuate a permanent relative birth. It is not

sufficient to define the living being as an organism. The living being is an organism from the point of view of its first individuation; but it can only live by being an organism that organizes and organizes itself in time; the organization of the organism is the result of a first individuation, which can be said to be absolute; but the latter is a condition of life rather than life itself; it is the condition of *the perpetual birth that is life itself*" (Simondon 1989, 171, translation and italics mine). Again, on the question of nonmetric time, see Massumi 2002, 185–86, 200.

5 ◻ LABORS OF REGENERATION: STEM CELLS AND THE EMBRYOID BODIES OF CAPITAL

Epigraph: Martin 2002, 3.

1. Here I am thinking of the labor theory of value as expounded in Marx's first volume of *Capital*. I have elsewhere argued that the labor theory of value is inseparable from a certain understanding of the fundamentals of human reproduction. See Cooper 2002. However, I also believe that another perspective on labor (productive and reproductive) can be extracted from Marx's work. In this sense I agree with such Italian autonomist thinkers as Antonio Negri ([1979] 1984), although the latter unduly neglects the sexual politics of labor and desire.

2. Certainly Marx believed that biological reproduction was subject to certain intrinsic limits, ranging from the natural rhythms of the seasons and the hours of the day to the germination and gestation times of plants and animals. As Marx ([1857] 1993, 742) nicely put it, there are inherent constraints on the reproducibility of meat: "In regard to the reproduction phase (especially circulation time), note that use value itself places limits upon it. Wheat must be reproduced in a year. Perishable things like milk etc. must be reproduced more often. Meat on the hoof does not need to be reproduced quite so often, since the animal is alive and hence resists time; but slaughtered meat on the market has to be reproduced in the form of money in the very short term, or it rots. The reproduction of value and of use value partly coincide, partly not."

3. See on this point Boyd and Watts 1997.

4. See in particular Hegel 1970 for an understanding of Hegel's conception of organic life. This work in particular seems to inform Marx's theory of human labor and reproduction in his first volume of *Capital*. However, there are parts of Marx's work that seem to evoke a decidedly more pathological—indeed monstrous—figure of animation.

5. On these quasi-cancerous properties, see Shostak 2001, 179–83.

6. On this point, see Parson 2004, 25–56. See also Cooper 2004.

7. This point is pursued furthest in terms of its conceptual and political consequences by the economist Christian Marazzi (2002). Marazzi thus avoids the twin dangers of a postmodern fetishism of the sign, on the one hand, and the nostalgic appeal to the fundamentals of production, on the other.

8. Thus, while I am entirely convinced by philosopher of science Hans-Jorg Rhein-

berger's characterization of scientific creativity as an encounter with the unforeseen consequences of the experimental process, I suggest that contemporary modes of capital accumulation are quite comfortable with the unexpected. The question of a counterpolitics of science thus becomes considerably more complex. See Rheinberger 1997.

9. See Geron 2003a.

10. On all these factors of uncertainty, see Geron 2003b.

11. While the biological patent can be situated in the longer history of patent right in general, it also marks a fundamental rupture—not only in the sense that it extends the scope of its coverage (life itself) but also because it reinvents the temporality of invention.

12. After establishing five unmodified human ES cell lines in 1998, Dr. James Thomson filed a patent application through the Wisconsin Alumni Research Foundation (WARF). U.S. Patent 6,200,806 was issued in 2001 and covers both the method of isolating human ES cells (the process) and the five unmodified stem cell lines themselves (the product). As the EU commission on stem cell patent comments, this "is a very broad patent right, which gives WARF control over *who may work* with these 5 human ES cell lines, over *who may use James Thomson's process* to isolate the stem cells, and over the *purpose* of the work (research, commercialisation)" (European Group on Ethics in Science and New Technologies to the European Commission 2002, 63). WARF then established an exclusive licensing agreement with Geron, which funded the research, while retaining the right to distribute its *unmodified* stem cell lines to academic researchers. Under this agreement Geron holds the rights to develop three *modified* types of stem cell line.

13. See A. Brimelow, "Inventions Involving Human Embryonic Stem Cells," April 2003. Available online on the Web site of the UK Intellectual Property Office, http://www.ipo.gov.uk/patent/p-decisionmaking/p-law/p-law-notice/p-law-notice-stemcells.htm.

14. The bill referenced is U.S. Congress, Consolidated Appropriations Act, 2004, HR 2673, Sec. 634, January 30, 2004.

15. On the problematization of traditional notions of human developmental potentiality in current stem cell research, see Waldby and Squier 2003. For an insight into the divergence between stem cell regeneration and the Weismannian model of germinal reproduction, see Cooper 2003. The patenting of human embryonic stem cells is the latest episode in a longer history of decisions on biological invention, dating back to 1980. In general, I suggest that two major points of disruption can be identified in the short history of the biotechnological patent, both of which have served to undermine the authority of Weismannian and legal conceptions of humanity: the first was the redefinition of life as transgenic (hence a whole series of test cases relating to recombinant DNA technologies and genetically engineered organisms throughout the 1980s); the second relates to recent developments in stem cell research and is particularly concerned with the legal status of the immortalized embryonic stem cell line. A decisive precedent for recent patents issued on human ES cell lines can be found in

the notorious John Moore case, where it was decided that an immortalized cancer cell line derived from the spleen of a patient could not be considered as his personal property and should instead be classified as a patentable invention. See on this point Lock 2002. The significance of ES cell patents, however, is much wider in the sense that it involves a wholesale redefinition of generation itself. What is at stake here is not merely a pathological, cancerous cell line, but the ultimate source of all bodily regeneration, according to stem cell scientists.

16. This patent is discussed in detail in the European Group on Ethics in Science and New Technologies to the European Commission 2002, 63.

17. For an illuminating perspective on the self-accumulative "bio-value" embodied in the stem cell line, see Waldby 2002. For cultural theorist Catherine Waldby the production of "bio-value" involves an intervention into the temporality of the body, one that attempts to modulate, arrest, or accelerate certain biological processes, with the ultimate aim of generating "a margin of bio-value, a surplus of fragmentary vitality" (ibid., 310).

18. Likewise, the most illuminating aspect of Marx's economics with regard to our present context is not his first volume of *Capital*, where he begins with and derives all value from the sphere of production, but rather the *Grundrisse* and the third volume of *Capital*, which instead work "backward" from the logic of speculative, financial capital.

19. On this point, see Marx [1894] 1981, 513, 607–10.

20. On this point I take issue with a pervasive tendency to interpret the growing commercializaton of the molecular and genetic life sciences as an instance of "commodification." See in particular the 2002 collection edited by Nancy Scheper-Hughes and Loïc Wacquant. I am not arguing here that the biological product has ceased to function as a commodity, but rather that the process of its commodification is now preceded by its transformation into speculative surplus value. In the current context of entrepreneurial science it is more important to own the speculative value of a cell line, through title to its "intellectual property," than to own the cell line itself. The property of the thing is included in and overwritten by the property of its future powers of emergence.

21. It is no accident then that developmental biologists are revisiting all kinds of biological enigmas in the history of the life sciences. In particular, self-regenerative animals such as the hydra (described as a permanent embryo) and teratocarcinomas (potentially immortal embryonal tumors). On this point, see Cooper 2003 and 2004. In Cooper 2002 I provide a more detailed reading of Marx's formula for capital and its relation to stem cell research.

22. Interestingly, even Marx was not immune from this prejudice, as the tone adopted in the third volume of his *Capital* makes clear. However, Marx's work is ambivalent enough to lend itself to other uses.

23. One exception here is the market in umbilical cord blood, which Waldby (2006) has analyzed as a form of emergent, biospeculative investment. Private cord

blood banks celebrate the potential consumer of these technologies as an entrepreneur of the self and a stakeholder in the future body of capital. Here, it seems, the biological life chances of the consumer are being incorporated into the same strategies of speculative life cycle investment explored by Randy Martin in his *Financialization of Daily Life* (2002).

24. For a preliminary discussion of the issues of reproductive labor, egg donation, and new reproductive technologies, see Dickenson 2001. Donna Dickenson makes the highly pertinent point that bioethical discussions of new reproductive and regenerative technologies routinely efface the whole question of women's reproductive labor, preferring to meditate on the dignity of the embryo.

25. However, for a first attempt to analyze the market in human eggs, see Waldby and Cooper 2007.

6 ❏ *THE UNBORN BORN AGAIN*

Epigraph: White House 2001.

1. I am here thinking of Brian Massumi's discussion of the temporal ellipsis in his "Requiem for Our Prospective Dead (Toward a Participatory Critique of Capitalist Power)" (1998). The motif of war was present in right-to-life rhetoric from the beginning. See, for example, Paul Marx's *The Death Peddlers: War on the Unborn, Past, Present, Future* (1971).

2. I here follow Nancy T. Ammerman's account of the American evangelical movement and its twentieth-century fundamentalist mutations. See Ammerman 1991. I am particularly concerned with the evangelical revival that occurred in the mid-1970s and has come to be associated with "born-againism" and pro-life politics. It might be argued that the born-again movement brings together the abiding concerns of the various evangelical strains of American Protestantism—republicanism, anti-authoritarianism, and personal rebirth—with the reactionary tendencies of Baptist fundamentalism. What is now known as the fundamentalist wing of evangelical Christianity emerged in the early part of the twentieth century as an internal reaction against progressive forces within the Protestant Church. "Fundamentalism," writes Ammerman, "differs from traditionalism or orthodoxy or even a mere revivalist movement. It differs in that it is a movement in conscious, organized opposition to the disruption of those traditions and orthodoxies" (ibid., 14). After fighting and losing battles to prohibit the teaching of evolution in schools, the fundamentalists retreated into relative political obscurity throughout the following decades, even as a new generation of nonseparatist evangelists such as Billy Graham were increasingly willing to engage in public and media life. It was only in the 1970s that this rift was repaired, as evangelicals started obsessing about the moral decline of America and the fundamentalists once again came out of hiding to do battle for their faith. No doubt this reunion accounts for the coexistence of apparently contradictory tendencies within the contemporary born-again movement: future-oriented, transformative, but

reactive nevertheless. The evangelical movement is generally understood to be an offshoot of mainline Protestantism. Other commentators have pointed out that both the Protestant and Catholic churches sprouted right-wing, evangelizing, and free-market wings around the same time. See, for example, Kintz 1997, 218, 226, 230. Certainly this convergence is evident in George W. Bush's frequent recourse to the advice of the Vatican. Because of this convergence, I cite the work of the Catholic free-market neoconservative Michael Novak (2001), who has had a considerable influence over (and arguably been influenced by) evangelical thinking.

3. However, there is a recent and growing literature on the role of emotions in finance. See in particular Pixley 2004. Two recent works are particularly interesting on the relationship between faith, credibility, credit/debt relations, and the question of political constitution. These are Aglietta and Orléan 1998 as well as Aglietta and Orléan 2002. Following the work of Michel Aglietta and André Orléan, I do not make any essential distinction between the gift and the debt, assuming that what constitutes a gift for one person will probably be experienced as a debt by another. Where I do draw a distinction is between different kinds and temporalities of the gift/debt relationship. In other words, the pertinent question here is whether or not the gift/debt is redeemable.

4. For an overview of Aquinas's economic philosophy, see the articles collected in Blaug 1991.

5. See, for example, Gilchrist 1969.

6. What interests me here is the importance of "born-againism" or regeneration within American evangelicalism *in general*. I make no attempt to provide an overview of the various denominational splits within American Protestant evangelicalism, although this would certainly be relevant for a historical understanding of the Republican–Southern Baptist alliance today. For a detailed insight into this history, see K. Phillips 2006.

7. There is thus an important distinction to be drawn between the Catholic philosophy of life (which presumes sovereign power) and the Protestant, evangelical culture of life (where life is understood as a form of self-regenerative debt). In the Protestant tradition sovereign power is not so much formative as *re*formative—it is the attempt to refound that which is without foundation. One important corollary of my argument is that Giorgio Agamben's philosophy of bare life (1998) is wholly unsuited to a critical engagement with the contemporary phenomenon of culture of life politics. Indeed, to the extent that he reinstates the sovereign model of power— if only in inverted form—as constitutive of power itself, his philosophical gesture comes very close to that of the right-to-life movement. Bare life, in other words, is the suspended inversion of the *vita beata* and finds its most popular iconic figure in the unborn fetus. Agamben's philosophy of biopolitics is not so much a negative theology as a theology in suspended animation.

8. For a complementary reading of U.S. debt and its role in the financialization of world capital markets, see Brenner 2002, 59ff, 206ff. See also Naylor 1987 for a fasci-

nating account of the links between neoliberalism, debt servitude, and neo-evangelical movements in South America and elsewhere. It should be noted here that not all contemporary evangelical philosophies of debt are necessarily imperialist. Liberation theology is one instance of a faith that works *against* third world debt.

9. The neoconservative movement is quite lucid about the speculative, future-oriented thrust of its return to fundamentals. It is here that one of the founding fathers of neoconservatism, Irving Kristol (1983, xii), identifies its distinguishing feature: "What is 'neo' ('new') about this conservatism," he proffers, "is that it is resolutely free of nostalgia. It, too, claims the future—and it is this claim, more than anything else, that drives its critics on the Left into something approaching a frenzy of denunciation."

10. Here I am thinking of philosopher Walter Benjamin's analysis of the cult in his 1921 essay "Capitalism as Religion" (see Benjamin 1999, 288–91). In this piece Benjamin asserts that the specificity of capitalism lies in its tendency to dispense with any specific dogma or theology other than the perpetuation of faith (ibid., 288). The religion of capital, he argues, comes into its own when God himself is included in the logic of the promise and can no longer function as its transcendent reference point or guarantor. In its ultimate cultic form the capitalist relation tends to become no more than a promise that sustains its own promise, a threat that sustains its own violence. The gifts it dispenses emanate from a promissory future and forego all anchorage in the past. In this sense it institutes a relation of guilt from which there is no relief or atonement.

11. There is some debate as to the intellectual sources of neoliberalism. In his recent history of the concept, the geographer David Harvey (2005, 54) discerns a complex fusion of monetarism, rational expectations, public choice theory, and the "less respectable but by no means uninfluential 'supply-side' ideas of Arthur Laffer." Like many others, he also points to the crucial role played by the journalist and investment analyst George Gilder in the popularization of neoliberal and supply-side economic ideas. However, here I follow economist Paul Krugman's more detailed analysis of supply-side theory to argue that the supply-siders actually offered a radical critique of neoclassical-inspired models of equilibrium economics such as monetarism. It was on the question of debt and budget deficits that at least some supply-siders took issue with the more traditional conservative economists. On these points, see Krugman 1994, 82–103, 151–69. The supply-side gospel has come to be associated with Reagonomics, and it was under Reagan that U.S. federal debt first began to outpace the GDP in relative terms. But by far the most extreme experiment in deficit free fall has been carried out under the administration of George W. Bush. Others have analyzed the religious dimension of neoliberalism by looking at Chicago-school monetarism. See, for example, Nelson 2002 and Taylor 2004. I tend to think that monetarism is an easy target and that supply-side ideas, particularly as espoused by Gilder, were much more influential on actual economic policy and popular cultures of neoliberalism. In this sense, too, I tend to see complexity-influenced approaches to

economics not as a counter to neoliberalism (as Taylor does) but as its ultimate expression. Gilder, for example, is a committed complexity theorist. For Gilder's thoughts on U.S. debt, see Gilder 1981, 230; and for his views on budget deficits under Bush, see Gilder 2004.

12. For a more detailed discussion of the sources of evangelical economics, see Lienesch 1993, 94–138.

13. On the history of *Roe v. Wade* and the Christian right, see Petchesky 1984. On the specific links between the right-to-life and born-again movements, see Harding 2000 183–209.

14. On this point, see Harding 2000, 189–91.

15. Again, feminist theorist Susan Friend Harding (ibid.) presents a compelling account of this identification in the work of fundamentalist Baptist Jerry Falwell. But it is recurrent in the literature of the period. For an insight into the born-again ethos of this era, see Graham 1979.

16. On the links between the right-to-life movement and white supremacist groups, see cultural theorist Carol Mason's astonishing essay "Minority Unborn" (1999, 159–74).

17. There is thus a fundamental ambivalence within the economic writings of the evangelicals, who on the one hand celebrate U.S. debt creationism and on the other obsess over the need to cancel all debt, restore strict tariff and exchange controls, and reinstate the gold standard. On this point, see Lienesch 1993, 104–7. Interestingly, the same ambivalence can be found among supply-side economists, some of whom advocate a return to the gold standard.

18. On the convergence of the neoconservatives and the religious right, see Diamond 1995, 178–202, as well as Halper and Clarke 2004, 196–200.

19. On the increasingly global reach of right-wing evangelical opinion, see Kaplan 2004, 219–43.

20. In his book *Holy Terrors: Thinking about Religion after September 11* (2003), the theorist of religion Bruce Lincoln explores the ways in which George W. Bush's speeches on foreign policy make implicit reference to the language of the religious right, often borrowing their syntax and phraseology from popular evangelical tracts on the apocalypse. I believe that a similar argument can be made with respect to Bush's pronouncements on the politics of life.

21. See again Mason 1999.

❏ *EPILOGUE*

1. David Harvey offers a preliminary analysis of what he calls "neoliberalism with Chinese characteristics" in his 2005 book *A Brief History of Neoliberalism*, 120–51.

2. On this point, see Greenhalgh and Winckler 2005.

References

Agamben, Giorgio. 1998. *Homo Sacer: Sovereign Power and Bare Life.* Translated by Daniel Heller-Roazen. Stanford, Calif.: Stanford University Press.

Aglietta, Michel, and Régis Breton. 2001. "Financial Systems, Corporate Control, and Capital Accumulation." *Economy and Society* 30, no. 4: 433–66.

Aglietta, Michel, and André Orléan. 2002. *La monnaie: Entre violence et confiance.* Paris: Odile Jacob.

———, eds. 1998. *La monnaie souveraine.* Paris: Odile Jacob.

Ammerman, Nancy T. 1991. "North American Protestant Fundamentalism." In *Fundamentalisms Observed.* Edited by Martin E. Marty and R. Scott Appleby, 1–63. Chicago: University of Chicago Press.

Anderson, Philip W., Kenneth J. Arrow, and David Pines, eds. 1988. *The Economy as an Evolving Complex System.* Redwood City, Calif.: Addison-Wesley.

Aquinas, Thomas. 1945. *Basic Writings of Saint Thomas Aquinas.* Volume 1. Edited by A. C. Pegis. New York: Random House.

Aradau, Claudia. 2004. "The Perverse Politics of Four-Letter Words: Risk and Pity in the Securitization of Human Trafficking." *Millennium: Journal of International Studies* 33, no. 2: 251–77.

Arnott, Jayne. 2004. "Sex Workers and Law Reform in South Africa." *HIV/ AIDS Law Policy Review* 9, no. 4.

Arrighi, Giovanni. 2002. "The African Crisis: World Systemic and Regional Aspects." *New Left Review* 15: 5–36.

———. 2003. "The Social and Political Economy of Global Turbulence." *New Left Review* 20: 5–71.

Arthur, Brian. W., Steven Durlauf, and David Lane, eds. 1997. *The Economy as an Evolving Complex System II.* Reading, Mass.: Addison-Wesley.

Ashcroft, Frances. 2001. *Life at the Extremes: The Science of Survival.* Hammersmith, London: Flamingo.

Ashforth, Adam. 2005. *Witchcraft, Violence, and Democracy in South Africa.* Chicago: University of Chicago Press.

Auger, Francois A., and Lucie Germain. 2004. "Tissue Engineering." In *Encyclopedia of Biomaterials and Biomedical Engineering.* Volume 2. Edited by Gary E. Wnek and Gary E. Bowlin, 1477–83. New York: Marcel Dekker.

Avant, Deborah D. 2005. *The Market for Force: The Consequences of Privatizing Security.* Cambridge: Cambridge University Press.

Bacher, Jamie M., Brian D. Reiss, and Andrew D. Ellington. 2002. "Anticipatory Evolution and DNA Shuffling." *Genome Biology* 3, no. 8: 1021–25.

Bak, Per. 1996. *How Nature Works: The Science of Self-Organized Complexity*. New York: Springer Verlag.

Bakker, Isabella. 2003. "Neo-liberal Governance and the Reprivatization of Social Reproduction: Social Provisioning and Shifting Gender Orders." In *Power, Production, and Social Reproduction: Human In/security in the Global Political Economy*. Edited by Isabella Bakker and Stephen Gill, 66–82. London: Palgrave Macmillan.

Bateson, William. 1992. *Materials for the Study of Variation Treated with Especial Regard to Discontinuity in the Origin of Species*. Baltimore, Md.: Johns Hopkins University Press.

Bauman, Zygmunt. 2004. *Wasted Lives*. Cambridge: Polity.

Bell, Daniel. 1974. *The Coming of Post-Industrial Society: A Venture in Social Forecasting*. London: Heinemann.

Benjamin, Walter. 1999. "Capitalism as Religion." In *Selected Writings, 1913–1926*. Edited by Marcus Bullock and Michael W. Jennings, 288–91. Cambridge, Mass.: Harvard University Press.

Bensaïd, Daniel. 1995. *La discordance des temps: Essais sur les crises, les classes, l'histoire*. Paris: Editions de la Passion.

Biggers, J. D. 1984. "In Vitro Fertilization and Embryo Transfer in Historical Perspective." In *In Vitro Fertilization and Embryo Transfer*. Edited by Alan Trounson and Carl Wood, 3–15. London: Churchill Livingstone.

Bigo, Didier. 2002. "Security and Immigration: Toward a Critique of the Governmentality of Unease." *Alternatives* 27: 63–92.

Billingham, R. E. 1976. "Concerning the Origins and Prospects of Cryobiology and Tissue Banking." *Transplantation Proceedings* 8, no. 2 (supplement 1) (June): 7–13.

Blaug, Mark, ed. 1991. *St. Thomas Aquinas (1225–1274), Pioneers in Economics*. Volume 3. Aldershot, England: Edward Elgar.

Block, Steven M. 1999. "Living Nightmares: Biological Threats Enabled by Molecular Biology." In *The New Terror: Facing the Threat of Biological and Chemical Weapons*. Edited by Sidney D. Drell, Abraham D. Sofaer, and George D. Wilson, 39–75. Stanford, Calif.: Hoover Institution Press.

Bond, Patrick. 2001. *Against Global Apartheid: South Africa Meets the World Bank, IMF, and International Finance*. Lansdowne, South Africa: University of Cape Town Press.

Borger, Julian. 2001. "Alarm as Bush Plans Health Cover for Unborn." *The Guardian*, July 7.

Bougen, Philip D. 2003. "Catastrophe Risk." *Economy and Society* 32, no. 2: 253–74.

Boutros-Ghali, Boutros. 1992. *An Agenda for Peace: Preventive Diplomacy, Peacemaking, and Peace-keeping*. New York: Department of Public Information, United Nations. Available online at http://www.un.org/Docs/SG/agpeace.html (accesssed March 2006).

Boyd, William, and Michael Watts. 1997. "Agri-industrial Just-in-Time." In *Globalising Food: Agrarian Questions and Global Restructuring*. Edited by David

Goodman and Michael Watts, 192–225. London: Routledge.

Boyer, Robert. 2000. "Is a Finance-led Growth Regime a Viable Alternative to Fordism? A Preliminary Analysis." *Economy and Society* 29, no. 1: 111–45.

Brenner, Robert. 2002. *The Boom and the Bubble: The U.S. in the World Economy*. London: Verso.

Brower, Jennifer, and Peter Chalk. 2003. *The Global Threat of New and Reemerging Infectious Diseases: Reconciling U.S. National Security and Public Health Policy*. Santa Monica, Calif.: RAND Corporation.

Buell, Frederick. 2003. *From Apocalypse to Way of Life: Environmental Crisis in the American Century*. London: Routledge.

Bush, George. 1997. Foreword. In Andrew S. Natsios, *U.S. Foreign Policy and the Four Horsemen of the Apocalypse: Humanitarian Relief in Complex Emergencies*. Xiii–xiv. Westport, Conn.: Praeger.

Buttel, Frederick H., Martin Kenney, and Jack Kloppenburg. 1985. "From Green Revolution to Biorevolution: Some Observations on the Changing Technological Bases of Economic Transformation in the Third World." *Economic Development and Cultural Change* 34, no. 1: 31–55.

Cache, Bernard. 1995. *Earth Moves: The Furnishing of Territories*. Translated by Anne Boyman. Cambridge, Mass.: MIT Press.

Caffentzis, George. 2006. "Acts of God and Enclosures in New Orleans." *Metamute Magazine*, May 24. Available online at http://www.metamute.org/en/node/7795/print (accessed October 2006).

Canguilhem, Georges. 1992. "Machine et organisme." In *La connaissance de la vie*, 124–59. Paris: Vrin.

Caplan, Arnold I. 2002. "In Vivo Remodelling." In *Reparative Medicine: Growing Tissues and Organs (Annals of the New York Academy of Science)*. Volume 961. Edited by J. D. Sipe, C. A. Kelley, and L. A. McNicol, 307–8. New York: New York Academy of Sciences.

Carr, Matt. 2005. "Energy Bill Boosts Industrial Biotechnology." *Industrial Biotechnology* (fall): 142–43.

Carrel, Alexis, and Charles A. Lindbergh. 1938. *The Culture of Organs*. London: Hamish Hamilton.

Carter, Ashton B. 2002. "The Architecture of Government in the Face of Terrorism." *International Security* 26, no. 3: 5–23.

Carter, Ashton B., and John P. White, eds. 2001. *Keeping the Edge: Managing Defense for the Future*. Cambridge, Mass.: MIT Press.

Chemical and Biological Arms Control Institute (CBACI) and Center for Strategic and International Studies (CSIS). 2000. *Contagion and Conflict: Health as a Global Security Challenge*. Washington, D.C.: CBACI / CSIS.

Chesnais, François, and Claude Serfati. 2000. "La gestion de l'innovation dans le régime d'accumulation à dominante financière." In *Connaissance et mondialisation*. Edited by Michel Delapierre, Philippe Moati, and El Mouhoub Mouhoud, 183–93. Paris: Economica.

Chichilnisky, Graciela, and Geoffrey Heal. 1998. "Economic Returns from the Biosphere." *Nature* 391: 629–30.

———. 1999. "Catastrophe Futures: Financial Markets for Unknown

Risks." In *Markets, Information, and Uncertainty*. Edited by Graciela Chichilnisky, 120–40. Cambridge: Cambridge University Press.

Chyba, Christopher. 1998. *Biological Terrorism, Emerging Diseases, and National Security*. New York: Rockefeller Brothers Fund.

———. 2000. *Conflict and Contagion: Health as a Global Security Challenge*. Washington, D.C.: CBACI / CSIS.

———. 2002. "Toward Biological Security." *Foreign Affairs* 81, no. 3: 122–36.

Clarke, Adele E. 1998. *Disciplining Reproduction: Modernity, American Life Sciences, and "the Problems of Sex."* Berkeley: University of California Press.

CNN. 2005. "Bush Military Bird Flu Role Slammed." *CNN*, October 5. Available online at http://edition.cnn.com/2005/POLITICS/10/05/bush.reax/ (accessed March 2006).

Cockell, Charles S. 2003. *Impossible Extinction: Natural Catastrophes and the Supremacy of the Microbial World*. Cambridge: Cambridge University Press.

Comaroff, Jean, and John L. Comaroff. 1993. *Modernity and Its Malcontents: Ritual and Power in Postcolonial Africa*. Chicago: University of Chicago Press.

———. 2001. "Millennial Capitalism: First Thoughts on a Second Coming." In *Millennial Capitalism and the Culture of Neoliberalism*. Edited by Jean Comaroff and John L. Comaroff, 1–56. Durham, N.C.: Duke University Press.

Cooper, Melinda. 2002. "The Living and the Dead: Variations on De Anima." *Angelaki: Journal of the Theoretical Humanities* 7, no. 3 (2002): 81–104.

———. 2003. "Rediscovering the Immortal Hydra: Stem Cells and the Question of Epigenesis." *Configurations* 11, no. 1: 1–26.

———. 2004. "Regenerative Medicine: Stem Cells and the Science of Monstrosity." *Medical Humanities* 30: 12–22.

———. 2007. "Marx Beyond Marx: A World Beyond and Outside Measure." In *Reading Negri*. Edited by Pierre Lamarche. London: Open Court Press.

Cordesman, Anthony H. 2001. *Terrorism, Asymmetric Warfare, and Weapons of Mass Destruction: Defending the U.S. Homeland*. Westport, Conn.: Praeger.

Coriat, Benjamin. 1994. *L'atelier et le chronomètre: Essai sur le taylorisme, le fordisme et la production de masse*. Paris: Christian Bourgois.

Coriat, Benjamin, and Fabienne Orsi. 2002. "Establishing a New Intellectual Property Rights Regime in the United States: Origins, Content, and Problems." *Research Policy* 31: 1491–507.

Correa, Gena. 1985. *The Mother Machine: Reproductive Technologies from Artificial Insemination to Artificial Wombs*. New York: Harper and Row.

Council of Environmental Quality and U.S. State Department. 1980. *The Global 2000 Report to the President of the U.S.: Entering the Twenty-first Century*. Volume 1, *The Summary Report*. New York: Pergamon.

Crush, Jonathan, and Wade Pendleton. 2004. *Regionalizing Xenophobia? Citizen Attitudes to Immigration and Refugee Policy in Southern Africa*. Cape Town, South Africa: Idasa.

Daily, Gretchen, and Katherine Ellison. 2002. *The New Economy of Nature—The Quest to Make Conservation Profitable.* Washington, D.C.: Island Press.

Davis, Mike. 1998. *Ecology of Fear: Los Angeles and the Imagination of Disaster.* New York: Random House.

———. 2005. *The Monster at Our Door: The Global Threat of Avian Flu.* New York: The New Press.

———. 2006. *Planet of Slums.* London: Verso.

DeLanda, Manuel. 2002. *Intensive Science and Virtual Philosophy.* New York: Continuum.

Deleuze, Gilles. 1990. *Logic of Sense.* Translated by M. Lester and C. Stivale. New York: Columbia University Press.

———. 1993. *The Fold: Leibniz and the Baroque.* Translated by Tom Conley. London: Athlone Press.

Diamond, Sara. 1995. *Roads to Dominion: Right-Wing Movements and Political Power in the United States.* New York: Guilford Press.

Dick, Steven J., and James E. Strick. 2004. *The Living Universe: NASA and the Development of Astrobiology.* New Brunswick, N.J.: Rutgers University Press.

Dickenson, Donna. 2001. "Property and Women's Alienation from Their Own Reproductive Labour." *Bioethics* 15, no. 3: 205–17.

Dickson, David. 1984. *The New Politics of Science.* New York: Pantheon Books.

Di Christina, Giuseppe, ed. 2001. *Architecture and Science.* London: Wiley-Academy.

D'Inverno, Mark, Neil D. Theise, and Jane Prophet. 2005. "Mathematical Modelling of Stem Cells: A Complexity Primer for the Stem Cell Biologist." In *Tissue Stem Cells: Biology and Applications.* Edited by Christopher Potten, Jim Watson, Robert Clarke, and Andrew Renehan, 1–16. New York: Marcel Dekker.

Dopfer, Kurt, ed. 2005. *The Evolutionary Foundations of Economics.* Cambridge: Cambridge University Press.

Drahos, Peter, and John Braithwaite. 2002. *Information Feudalism: Who Owns the Knowledge Economy?* London: Earthscan Publications.

Dubos, René. [1959] 1987. *Mirage of Health: Utopias, Progress, and Biological Change.* Reprint, New Brunswick, N.J.: Rutgers University Press.

———. 1961. "Integrative and Creative Aspects of Infection." In *Perspectives in Virology.* Edited by M. Pollard, 200–5. Minneapolis, Minn.: Burgess.

Edelman, Bernard. 1989. "Le droit et le vivant." *La recherche* 20: 966–76.

Edwards, R. G. 2001. "IVF and the History of Stem Cells." *Nature* 413: 349–51.

Elbe, Stefan. 2002. "HIV/AIDS and the Changing Landscape of War in Africa." *International Security* 27, no. 2: 159–77.

———. 2005. "AIDS, Security, Biopolitics." *International Relations* 19, no. 4: 403–19.

Esposito, Roberto. 2002. *Immunitas: Protezione e negazione della vita.* Turin: Einaudi.

Estes, Carroll L., and associates. 2001. *Social Policy and Aging: A Critical Perspective.* Thousand Oaks, Calif.: Sage Publications.

European Group on Ethics in Science

and New Technologies to the European Commission. 2002. "Opinion on Ethical Aspects of Patenting Inventions Involving Human Stem Cell Research." Luxembourg: Office for Official Publications of the European Commission.

Ewald, François. 1986. *L'etat providence*. Paris: Grasset et Fasquelle.

———. 1993. "Two Infinities of Risk." In *The Politics of Everyday Fear*. Edited by Brian Massumi, 221–28. Minneapolis: University of Minnesota Press.

———. 2002. "The Return of Descartes's Malicious Demon: An Outline of a Philosophy of Precaution." In *Embracing Risk: The Changing Culture of Insurance and Responsibility*. Edited by Tom Baker and Jonathon Simon, 273–301. Chicago: University of Chicago Press.

Ewald, Paul. 2002. *Plague Time: The New Germ Theory of Disease*. New York: Anchor.

"Excerpts from Pentagon's Plan: 'Prevent the Re-Emergence of a Rival.'" 1992. *New York Times*, March 8.

Ferguson, James. 2006. *Global Shadows: Africa in the Neoliberal World Order*. Durham, N.C.: Duke University Press.

Fletcher, Liz. 2001. "Re-engineering the Business of Regenerative Medicine." *Nature Biotechnology* 19: 204–5.

Fortun, Michael. 2001. "Mediated Speculations in the Genomics Futures Markets." *New Genetics and Society* 20, no. 2: 139–56.

Foucault, Michel. 1973. *The Order of Things: An Archaeology of the Human Sciences*. New York: Vintage Books.

———. 2003. *"Society Must Be Defended": Lectures at the Collège de France, 1975–1976*. Translated by D. Macey. London: Allen and Unwin.

———. 2004. *La naissance de la biopolitique: Cours au Collège de France 1978–1979*. Paris: Gallimard / Seuil.

Franklin, Sarah. 2006. "Embryonic Economies: The Double Reproductive Value of Stem Cells." *Biosocieties* 1: 71–90.

Fraser, Claire M., and Malcolm R. Dando. 2001. "Genomics and Future Biological Weapons: The Need for Preventative Action by the Biomedical Community." *Nature Genetics* 29: 253–65.

Fraumann, Edwin. 1997. "Economic Espionage: Security Missions Redefined." *Public Administration Review* 57, no. 4 (July–August): 303–8.

Freed, Lisa E., and Gordana Vunjak-Novakovic. 2000. "Tissue Engineering Bioreactors." In *Principles of Tissue Engineering*. 2nd edition. Edited by Robert P. Lanza, Robert Langer, and William L. Chick, 143–56. San Diego, Calif.: Academic Press.

Gadrey, Jean. 2003. *New Economy, New Myth*. London: Routledge.

Gardels, Nathan. 2002. "Why Not Preempt Global Warming?" *New Perspectives Quarterly* (fall): 2–3.

George, Susan, and Fabrizio Sabelli. 1994. *Faith and Credit: The World Bank's Secular Empire*. London: Penguin.

Geron Corporation. 2000. *Annual Report*. Available online at http://www.geron.com/annualreports/GeronAnnualReport2000.pdf (accessed October 2005).

———. 2003a. *Annual Report*. Available online at http://www.geron.com/annualreports/GeronAnnualReport2003.pdf (accessed October 2005).

———. 2003b. "Form 10-Q for Geron Corporation." *Quarterly Report*, 12 November.

Gilchrist, J. 1969. *The Church and Economic Activity in the Middle Ages*. London: Macmillan.

Gilder, George. 1981. *Wealth and Poverty*. New York: Basic Books.

———. 1986. *Men and Marriage*. New York: Basic Books.

———. 2004. "Market Economics and the Conservative Movement." *Philadelphia Society Address*, June 1. Available online at http://www.discovery.org/scripts/viewDB/index.php?command=view&id=2061 (accessed March 2006).

Goodwin, Brian, and Gerry Webster. 1996. *Form and Transformation: Generative and Relational Principles in Biology*. New York: Cambridge University Press.

Goozner, Merrill. 2006. "Can Government Go Green?" *American Prospect Online*, March 19. Available online at http://www.prospect.org/cs/articles?article=can_government_go_green (accessed March 2006).

Graham, Billy. 1979. *The Holy Spirit: Activating God's Power in Your Life*. London: Collins.

Green, Ronald M. 2001. *The Human Embryo Research Debates: Bioethics in the Vortex of Controversy*. New York: Oxford University Press.

Greenhalgh, Susan, and Edwin A. Winckler. 2005. *Governing China's Population: From Leninist to Neoliberal Biopolitics*. Stanford, Calif.: Stanford University Press.

Guattari, Félix. 1995. *Chaosophy*. Edited by Sylvère Lotringer. New York: Semiotext(e).

Guillemin, Jeanne. 2004. *Biological Weapons: From the Invention of State-Sponsored Programs to Contemporary Bioterrorism*. New York: Columbia University Press.

Hajer, Maarten A. 1995. *The Politics of Environmental Discourse: Ecological Modernization and the Policy Process*. Oxford: Clarendon Press.

Hall, Molly J., Ann E. Norwood, Robert J. Ursano, and Carol S. Fullerton. 2003. "The Psychological Impacts of Bioterrorism." *Biosecurity and Bioterrorism: Biodefense Strategy, Practice, and Science* 1, no. 2: 139–44.

Haller, Stephen F. 2002. *Apocalypse Soon? Wagerings on Warnings of Global Catastrophe*. Montreal: McGill-Queen's University Press.

Halper, Stefan, and Jonathan Clarke. 2004. *America Alone: The Neo-Conservatives and the Global Order*. Cambridge: Cambridge University Press.

Hammond, Edward. 2001–2. "Profits of Doom." *The Ecologist* 31, no. 10: 42–5.

Haour-Knipe, M., and R. Rector, eds. 1996. *Crossing Borders: Migration, Ethnicity, and AIDS*. London: Taylor and Francis.

Harding, Susan Friend. 2000. *The Book of Jerry Falwell: Fundamentalist Language and Politics*. Princeton, N.J.: Princeton University Press.

Harris, Geoff. 2002. "The Irrationality of South Africa's Military Expenditure."

African Security Review 11, no. 2. Available online at http://www.iss .co.za/Pubs/ASR/11No2/Harris.html (accessed April 2006).

Harvey, David. 2005. *A Brief History of Neoliberalism*. Oxford: Oxford University Press.

Hawken, Paul, Amory B. Lovins, and L. Hunter Lovins. 1999. *Natural Capitalism: The Next Industrial Revolution*. London: Earthscan.

Hayek, Friedrich von. 1969. *Studies in Philosophy, Economics, and Politics*. New York: Simon and Schuster.

Healy, Gene. 2003. "Deployed in the U.S.A: The Creeping Militarization of the Home Front." *Cato Institute Policy Analysis*, no. 503 (December 17). Available online at http://www.cato .org/pubs/pas/pa-503es.html.

———. 2005. "Domestic Militarization: A Disaster in the Making." *Cato Institute*, September 27. Available online at http://www.cato.org/pub_display.php? pub_id=5074&print=Y (accessed March 2006).

Hegel, G. W. F. 1970. *Hegel's Philosophy of Nature*. Translated and edited by M. J. Petry. London: Allen and Unwin.

Heidegger Martin. [1938] 1977. "The Age of the World Picture." In Martin Heidegger, *The Question Concerning Technology and Other Essays*. Translated by William Lovitt, 115–45. New York: Harper Row.

Helmreich, Stefan. 2001. "Artificial Life, INC.: Darwin and Commodity Fetishism from Santa Fe to Silicon Valley." *Science as Culture* 10, no. 4: 483–504.

Henwood, Doug. 2003. *After the New Economy*. New York: The New Press.

Ho, Mae-Wan. 1999. *Genetic Engineering: Dream or Nightmare?* Dublin: Gateway.

Hudson, Michael. 2003. *Super Imperialism: The Origin and Fundamentals of U.S. World Dominance*. London: Pluto.

———. 2005. *Global Fracture: The New International Economic Order*. London: Pluto Press.

Ingber, Donald E. 2003. "Tensegrity II: How Structural Networks Influence Cellular Information Processing Networks." *Journal of Cell Science* 116, no. 8: 1397–408.

Jasanoff, Sheila. 2005. *Designs on Nature: Science and Democracy in Europe and the United States*. Princeton, N.J.: Princeton University Press.

Jessop, Bob. 2002. *The Future of the Capitalist State*. Cambridge: Polity.

Johnson, Loch K., and Diane C. Snyder. 2001. "Beyond the Traditional Intelligence Agenda: Examining the Merits of a Global Public Health Portfolio." In *Plagues and Politics: Infectious Disease and International Policy*. Edited by Andrew T. Price-Smith, 214–33. London: Palgrave.

Jordanova, Ludmilla. 1995. "Interrogating the Concept of Reproduction in the Eighteenth Century." In *Conceiving the New World Order*. Edited by Faye D. Ginsburg and Rayna Rapp, 369–86. Berkeley: University of California Press.

Kaplan, Esther. 2004. *With God on Their Side: How Christian Fundamentalists Trampled Science, Policy, and Democracy in George W. Bush's White House*. New York: The New Press.

Kauffman, Stuart. 1995. *At Home in the Universe*. Oxford: Oxford University Press.

———. 2000. *Investigations*. Oxford: Oxford University Press.

Kempadoo, Kamala, and Jo Doezema, eds. 1998. *Global Sex Workers: Rights, Resistance, and Redefinition*. New York: Routledge.

Kenney, Martin. 1986. *Biotechnology: The University-Industrial Complex*. New Haven, Conn.: Yale University Press.

Kintz, Linda. 1997. *Between Jesus and the Market: The Emotions That Matter in Right-Wing America*. Durham, N.C.: Duke University Press.

Kloppenburg, Jack Ralph. 1988. *First the Seed: The Political Economy of Plant Biotechnology, 1492–2000*. Cambridge: Cambridge University Press.

Kolnai, Aurel. [1929] 2004. "Disgust." In *On Disgust*. Edited by Barry Smith and Carolyn Korsmeyer, 29–91. Chicago: Open Court Press.

Kristol, Irving. 1983. *Reflections of a Neoconservative: Looking Back, Looking Ahead*. New York: Basic Books.

Kristol, William, and George Weigel. 1994. "Life and the Party." *National Review* 15: 53–7.

Krugman, Paul. 1994. *Peddling Prosperity: Economic Sense and Nonsense in the Age of Diminished Expectations*. New York: Norton.

Landecker, Hannah. 2005. "Living Differently in Time: Plasticity, Temporality, and Cellular Biotechnologies." *Culture Machine* (biopolitics issue). Available online at http://culturemachine.tees.ac.uk/frm_f1.htm (accessed March 2006).

Lederberg, Joshua, Robert E. Shope, and Stanley C. Oaks, eds. 1992. *Emerging Infections: Microbial Threats to Health in the United States*. Washington, D.C.: National Academy Press.

Lefters, Llewellyn, Linda Brink, and Ernest Takafuji. 1993. "Are We Prepared for a Viral Epidemic Emergency?" In *Emerging Viruses*. Edited by Stephen Morse. New York: Oxford University Press, 272.

Lehrman, Lewis E. 1986. "The Right to Life and the Restoration of the American Republic." *National Review* 38: 25–30.

Leitenberg, Milton, James Leonard, and Richard Spertzel. 2004. "Biodefense Crossing the Line." *Politics and the Life Sciences* 22, no. 2: 1–2.

Le Méhauté, Alain. 1990. *Fractal Geometries: Theory and Applications*. Translated by Jack Howlett. London: Penton Press.

Lerner, Eric J. 1991. *The Big Bang Never Happened*. New York: Random House.

Levy, S. B, and R. P. Novick, eds. 1986. *Antibiotic Resistance Genes: Ecology, Transfer, and Expression*. New York: Cold Spring Harbor Laboratory.

Lienesch, Michael. 1993. *Redeeming America: Piety and Politics in the New Christian Right*. Chapel Hill: University of North Carolina Press.

Lincoln, Bruce. 2003. *Holy Terrors: Thinking about Religion after September 11*. Chicago: University of Chicago Press.

Lock, Margaret. 2001. *Twice Dead: Organ Transplants and the Reinvention of Death*. Berkeley: University of California Press.

———. 2002. "The Alienation of Body

Tissue and the Biopolitics of Immortalized Cell Lines." In *Commodifying Bodies*. Edited by Nancy Scheper-Hughes and Loïc Wacquant, 63–92. London: Sage.

Loeppky, Rodney. 2005. *Encoding Capital: The Political Economy of the Human Genome Project*. New York: Routledge.

Longmore, Donald. 1968. *Spare-Part Surgery: The Surgical Practice of the Future*. London: Aldus Books.

Lovelock, James E. 1987. *Gaia: A New Look at Life on Earth*. Oxford: Oxford University Press.

Luxemburg, Rosa. 1973. *The Accumulation of Capital—An Anti-Critique (with Imperialism and the Accumulation of Capital, by Nikolai I. Bukharin)*. Edited and with an introduction by Kenneth J. Tarbuck. Translated by Rudolf Wichmann. New York: Monthly Review Press.

Lynn, Gregg. 1999. *Animate Form*. New York: Princeton Architectural Press.

Maienschein, Jane. 2003. *Whose View of Life? Embryos, Cloning, and Stem Cells*. Cambridge, Mass.: Harvard University Press.

Mampaey, Luc, and Claude Serfati. 2004. "Les groupes d'armement et les marchés financiers: Vers une convention 'guerre sans limites'?" In *La finance mondialisée: Racines sociales et politiques, configuration, conséquences*. Edited by François Chesnais, 223–51. Paris: La Découverte.

Mandelbrot, Benoit. 2004. *The (Mis)Behavior of Markets: A Fractal View of Risk, Ruin, and Reward*. London: Profile Books.

Marazzi, Christian. 2002. *Capitale e linguaggio: Dalla new economy all'economia di guerra*. Rome: DeriveApprodi.

Margulis, Lynn. 2004. "Gaia by Any Other Name." In *Scientists Debate Gaia: The Next Century*. Edited by Stephen H. Schneider, James R. Miller, Eileen Crist, and Penelope J. Boston, 7–12. Cambridge, Mass.: MIT Press.

Margulis, Lynn, and Dorion Sagan. 1997. *Microcosmos: Four Billion Years of Evolution from Our Microbial Ancestors*. Berkeley: University of California Press.

Marshall, F. H. A. 1910. *Physiology of Reproduction*. London: Longmans Green.

Martin, Randy. 2002. *Financialization of Daily Life*. Philadelphia: Temple University Press.

Marx, Karl. [1857] 1993. *Grundrisse: Foundations of the Critique of Political Economy*. Translated by Martin Nicolaus. Harmondsworth, England: Penguin.

———. [1867] 1990. *Capital*. Volume 1. Translated by Ben Fowkes. Harmondsworth, England: Penguin.

———. [1894] 1981. *Capital: A Critique of Political Economy*. Volume 3. Translated by David Fernbach. Harmondsworth, England: Penguin.

Marx, Paul. 1971. *The Death Peddlers: War on the Unborn, Past, Present, Future*. Collegeville, Minn.: Human Life International.

Mason, Carol. 1999. "Minority Unborn." In *Fetal Subjects, Feminist Positions*. Edited by Lynn M. Morgan and Meredith M. Michaels, 159–74. Philadelphia: University of Pennsylvania Press.

Massumi, Brian. 1998. "Requiem for Our Prospective Dead (Toward a Participatory Critique of Capitalist Power)." In

Deleuze and Guattari: New Mappings in Politics, Philosophy, and Culture. Edited by Eleanor Kaufman and Kevin Jon Heller, 40–64. Minneapolis, Minn.: University of Minnesota Press.

———. 2002. *Parables for the Virtual: Movement, Affect, Sensation.* Durham, N.C.: Duke University Press.

McCoy, Alfred W. 2003. *The Politics of Heroin: CIA Complicity in the Global Drug Trade.* Chicago: Lawrence Hill Books.

Meadows, Donella H., Dennis L. Meadows, and Jorgen Randers. 1992. *Beyond the Limits: Global Collapse or a Sustainable Future.* London: Earthscan Publications.

Meadows, Donella H., Dennis L. Meadows, Jorgen Randers, and William W. Behrens. 1972. *The Limits to Growth: A Report for the Club of Rome's Project on the Predicament of Mankind.* London: Pan Books.

Mejia, Lito C., and Kent S. Vilendrer. 2004. "Bioreactors." In *Encyclopedia of Biomaterials and Biomedical Engineering.* Volume 1. Edited by Gary E. Wnek and Gary L. Bowlin, 103–20. New York: Marcel Dekker.

Miller, Judith, Stephen Engelberg, and William Broad. 2001. *Germs: The Ultimate Weapon.* New York: Simon and Schuster.

Miller, Robert V., and Martin J. Day. 2004. "Horizontal Gene Transfer and the Real World." In *Microbial Evolution: Gene Establishment, Survival, and Exchange.* Edited by R. V. Miller and M. J. Day. Washington, D.C: American Society of Microbiology Press.

Minot, Charles S. 1908. *The Problem of Age, Growth, and Death.* London: John Murray.

Mirowski, Philip. 1996. "Do You Know the Way to Santa Fe?" In *New Directions in Political Economy: Malvern after Ten Years.* Edited by Steve Pressman, 13–40. London: Routledge.

———. 1997. "Machine Dreams: Economic Agents as Cyborgs." In *New Economics and Its History.* Edited by John B. Davis, 13–40. Durham, N.C.: Duke University Press.

Müller, Harald, and Mitchell Reiss. 1995. "Counterproliferation: Putting New Wine in Old Bottles." In *Weapons Proliferation in the 1990s.* Edited by Brad Roberts, 139–50. Cambridge, Mass.: MIT Press.

Nancy, Jean-Luc. 2002. *La création du monde ou la mondialisation.* Paris: Galilée.

National Intelligence Council. 2000. *National Intelligence Estimate: The Global Infectious Disease Threat and Its Implications for the United States.* Washington, D.C.: National Intelligence Council. January. Available online at http://www.ciaonet.org/wps/nic01/nic01.pdf (accessed March 2006).

National Security Strategy. 2002. "The National Security Strategy of the United States of America." September. Available online at http://www.white house.gov/nsc/nss.html (accessed March 2006).

Natsios, Andrew S. 1997. *U.S. Foreign Policy and the Four Horsemen of the Apocalypse: Humanitarian Relief in Complex Emergencies.* With foreword by George Bush. Westport, Conn.: Praeger.

Naughton, Gail K. 2002. "From Lab Bench to Market: Critical Issues in Tissue Engineering." In *Reparative*

Medicine: Growing Tissues and Organs (*Annals of the New York Academy of Science, vol. 961*). Edited by J. D. Sipe, C. A. Kelley. and L. A. McNicol, 372–85. New York: New York Academy of Sciences.

Naylor, R. T. 1987. *Hot Money and the Politics of Debt*. London: Unwin Hyman.

Negri, Antonio. [1979] 1984. *Marx Beyond Marx: Lessons on the Grundrisse*. Translated by Harry Cleaver, Michael Ryan, and Maurizio Viano. Edited by Jim Fleming. Boston, Mass.: Bergin and Garvey Publishers.

Negri, Antonio, and Michael Hardt. 2001. *Empire*. Cambridge, Mass.: Harvard University Press.

———. 2004. *Multitude: War and Democracy in the Age of Empire*. New York: Penguin Press.

Nelson, Robert H. 2002. *Economics as Religion: From Samuelson to Chicago and Beyond*. University Park, Pa.: Penn State University Press.

Noll, Mark A. 2002. *America's God: From Jonathan Edwards to Abraham Lincoln*. New York: Oxford University Press.

Novak, Michael. 2001. "The Principle's the Thing: On George Bush and Embryonic Stem-cell Research." *National Review Online*, August 10. Available online at http://www.nationalreview.com/contributors/novakprint081001.html (accessed March 2006).

Novick, Richard, and Seth Shulman. 1990. "New Forms of Biological Warfare?" In *Preventing a Biological Arms Race*. Edited by Susan Wright, 103–19. Cambridge, Mass.: MIT Press.

Office of Force Transformation, U.S. Department of Defense. 2004.

Elements of Defense Transformation. Washington, D.C.: U.S. Department of Defense. October. Available online at http://www.oft.osd.mil/library/library_files/document_383_ElementsOf Transformation_LR.pdf (accessed March 2006).

O'Hanlon, Michael E., Susan E. Rice, and James B. Steinberg. 2002. "The New National Security Strategy and Preemption (Policy Brief 113)." Washington, D.C.: Brookings Institution, December. Available online at http://www.brook.edu/comm/policybriefs/pb113.htm (accessed March 2006).

Organisation for Economic Cooperation and Development (OECD). 2004. *Biotechnology for Sustainable Growth and Development*. Paris: OECD Publications.

———. 2005. *Proposal for a Major Project on the Bioeconomy in 2030*. Paris: OECD Publications.

Parson, Ann B. 2004. *Proteus Effect: Stem Cells and Their Promise for Medicine*. Washington, D.C.: Joseph Henry Press.

Peberdy, Sally, and Natalya Dinat. 2005. *Migration and Domestic Workers: Worlds of Work, Health, and Mobility in Johannesburg*. Cape Town, South Africa: Idasa.

Pederson, Roger A. 1999. "Ethics and Embryonic Cells." *Scientific American* 280, no. 1 (April): 47.

Petchesky, Rosalind P. 1984. *Abortion and Woman's Choice: The State, Sexuality, and Reproductive Freedom*. New York: Longman.

Phillips, Kevin. 2004. *American Dynasty: Aristocracy, Fortune, and the Politics of Deceit in the House of Bush*. London: Penguin Books.

———. 2006. *American Theocracy: The*

Perils and Politics of Radical Religion, Oil, and Borrowed Money in the Twenty-first Century. New York: Viking.

Phillips, Michael G. 1991. *Organ Procurement, Preservation, and Distribution in Transplantation.* Richmond, Va.: United Network for Organ Sharing.

Pignarre, Philippe. 2003. *Le grand secret de l'industrie pharmaceutique.* Paris: La Découverte.

Pirages, Dennis, and Ken Cousins, eds. 2005. *From Resource Scarcity to Ecological Security: Exploring New Limits to Growth.* Cambridge, Mass.: MIT Press.

Pixley, Jocelyn. 2004. *Emotions in Finance: Distrust and Uncertainty in Global Markets.* Cambridge: Cambridge University Press.

Prigogine, Ilya, and Dilip K. Kondepudi. 1998. *Modern Thermodynamics: From Heat Engines to Dissipative Structures.* New York: John Wiley.

Prigogine, Ilya, and Grégoire Nicolis. 1989. *Exploring Complexity: An Introduction.* New York: W. H. Freeman.

Prigogine, Ilya, and Isabelle Stengers. 1979. *La nouvelle alliance: Métamorphose de la science.* Paris: Gallimard.

———. 1984. *Order out of Chaos: Man's New Dialogue with Nature.* London: Heinemann.

———. 1992. *Entre le temps et l'éternité.* Paris: Flammarion.

Rabinbach, Anson. 1992. *The Human Motor: Energy, Fatigue, and the Origins of Modernity.* Berkeley: University of California Press.

Razvi, Enal S., and Jonathan Burbaum. 2006. *Life Science Mergers and Acquisitions.* Westborough, Mass.: Drug & Market Development Publications.

Rensberger, Boyce. 1996. *Life Itself:*

Exploring the Realm of the Living Cell. Oxford: Oxford University Press.

Resnick, David P. 2002. "Bioterrorism and Patent Rights: 'Compulsory Licensure' and the Case of Cipro." *American Journal of Bioethics* 2, no. 3: 29–39.

Reuleaux, Franz. [1875] 1963. *The Kinematics of Machinery: Outline of a Theory of Machines.* Translated by Alexander Kennedy. Reprint, New York: Dover.

Rheinberger, Hans-Jorg. 1997. "Experimental Complexity in Biology: Some Epistemological and Historical Remarks." *Philosophy of Science* 64, supplement (December): S245–S254.

Rist, Gilbert. 2004. *The History of Development from Western Origins to Global Faith.* London: Zed Books.

Robertson, Pat. 1991. *The New World Order.* Dallas, Tex.: Word Publishing.

Ross, Andrew. 1991. *Strange Weather: Culture, Science, and Technology in the Age of Limits.* London: Verso.

Rostow, Walt W. 1960. *The Stages of Economic Growth: A Non-Communist Manifesto.* Cambridge: Cambridge University Press.

Rothschild, Emma. 1995. "What Is Security?" *Daedalus* 124 (summer): 53–98.

Rothschild, Lynn J., and Rocco L. Mancinelli. 2001. "Life in Extreme Environments." *Nature* 409: 1092–101.

Rubinsky, Boris. 2002. "Low Temperature Preservation of Biological Organs and Tissues." In *Future Strategies for Tissue and Organ Replacement.* Edited by Julia M. Polak, Larry L. Hench, and P. Kemp, 27–49. London: Imperial College Press.

Sapp, Jan. 2003. *Genesis: The Evolution of*

Biology. New York: Oxford University Press.

Sassen, Saskia. 2003. "Global Cities and Survival Circuits." In *Global Woman: Nannies, Maids, and Sex Workers in the New Economy.* Edited by Barbara Ehrenreich and Arlie Russell Hothschild, 254–74. London: Granta Books.

Saunders, Penelope. 2005. "Traffic Violations: Determining the Meaning of Violence in Sexual Trafficking versus Sex Work." *Journal of Interpersonal Violence* 20, no. 3 (March): 343–60.

Scheper-Hughes, Nancy, and Loïc Wacquant, eds. 2002. *Commodifying Bodies.* London: Sage.

Schleifer, R. 2005. "United States: Challenges Filed to Anti-Prostitution Pledge Requirement." *HIV/AIDS Policy Law Review* 10, no. 3: 21–23.

Schmitt, Carl. [1950] 2003. *Nomos of the Earth in the International Law of Ius Publicum Europaeum.* Translated by G. L. Ulmen. New York: Telos Press.

Schumpeter, Joseph A. 1934. *The Theory of Economic Development: An Inquiry into Profits, Capital, Credit, Interest, and the Business Cycle.* Translated by Redvers Opie. Cambridge, Mass.: Harvard University Press.

Schwartz, Peter, and Doug Randal. 2003. "An Abrupt Climate Change Scenario and Its Implications for United States National Security." Available online at http://www.gbn.com/ArticleDisplay Servlet.srv?aid=26231 (accessed March 2006).

Sell, Susan K. 2003. *Private Power, Public Law: The Globalization of Intellectual Property Rights.* Cambridge: Cambridge University Press.

Shostak, Stanley. 2001. *Becoming Immor-*

tal: Combining Cloning and Stem-Cell Therapy. New York: State University of New York Press.

Simmel, Georg. 1978. *The Philosophy of Money.* London: Routledge and Kegan Paul.

Simon, Julian L. 1996. *The Ultimate Resource 2.* Princeton, N.J.: Princeton University Press.

Simon, Julian L., and Herman Kahn. 1984. *The Resourceful Earth: A Response to Global 2000.* Oxford: Blackwell.

Simondon, Gilbert. 1989. *L'individuation psychique et collective.* Paris: Aubier.

———. 1995. *L'individu et sa genèse physico-biologique.* Grenoble, France: Jérôme Millon.

Singer, Peter F. 2003. *Corporate Warriors: The Rise of the Privatized Military Industry.* Ithaca, N.Y.: Cornell University Press.

Sitze, Adam. 2004. "Denialism." *South Atlantic Quarterly* 103, no. 4: 769–811.

Slater, Dashka. 2002. "Humouse." *Legal Affairs* (November–December). Available online at http://www.legalaffairs .org/issues/November-December-2002/feature_slater_novdec2002.html (accessed October 2005).

Slaughter, Barbara. 1999. "Cape Town Promotes Sex Tourism." *World Socialist Website.* Available online at http:// www.wsws.org/articles/1999/oct1999/ saf-o05_prn.shtml (accessed March 2006).

Sontag, Susan. 1988. *AIDS and Its Metaphors.* New York: Farrar, Straus and Giroux.

Stocum, David L. 1998. "Bridging the Gap: Restoration of Structure and Function in Humans." In *Cellular and Molecular Basis of Regeneration:*

From Invertebrates to Humans. Edited by Patrizia Ferretti and Jacqueline Géraudie. New York: Wiley.

Stoler, Laura. 1995. *Race and the Education of Desire: Foucault's History of Sexuality and the Colonial Order of Things.* Durham, N.C.: Duke University Press.

Strathern, Marilyn. 1999. "Potential Property: Intellectual Rights and Property in Persons." *Property, Substance, and Effect: Anthropological Essays on Persons and Things.* 161–77. London: Athlone Press.

Sun, Wei, Andrew Darling, Binil Starly, and Jae Nam. 2004. "Computer-aided Tissue Engineering: Overview, Scope, and Challenges (Review)." *Biotechnological Applications in Biochemistry* 39: 29–47.

Sun, Wei, Binil Starly, Andrew Darling, and Connie Gomez. 2004. "Computer-aided Tissue Engineering: Application to Biomimetic Modelling and Design of Tissue Scaffolds." *Biotechnological Applications in Biochemistry* 39: 49–58.

Sunder Rajan, Kaushik. 2006. *Biocapital: The Constitution of Post-Genomic Life.* Durham, N.C.: Duke University Press.

Swiss Re. 1998. "Genetic Engineering and Liability Insurance: The Power of Public Perception." Available online at http://www.swissre.com/INTERNET/ pwsfilpr.nsf/vwFilebyIDKEYLu/WW IN-4VFDC7/$FILE/genetic_eng.Paras .0003.File.pdf (accessed March 2006).

Taylor, Mark C. 2004. *Confidence Games: Money and Markets in a World without Redemption.* Chicago: University of Chicago Press.

Thom, René. 1975. *Structural Stability and Morphogenesis: An Outline of a General Theory of Models.* Translated by D. H. Fowler. Reading, Mass.: W. A. Benjamin Inc.

Thompson, Charis. 2005. *Making Parents: The Ontological Choreography of Reproductive Technologies.* Cambridge, Mass.: MIT Press.

Thompson, D'Arcy Wentworth. [1917] 1992. *On Growth and Form (An Abridged Edition).* Cambridge: Cambridge University Press.

Titmuss, Richard M. 1971. *The Gift Relationship: From Human Blood to Social Policy.* New York: Pantheon Books.

Townsend, Mark, and Paul Harris. 2004. "Now the Pentagon Tells Bush: Climate Change Will Destroy Us." *The Observer*, February 22. Available online at http://www.guardian.co .uk/print/0,3858,4864237–110970,00. html (accessed March 2006).

Travis, John. 2003. "Interview with Michael Goldblatt, Director, Defense Sciences Office, DARPA." *Biosecurity and Bioterrorism: Biodefense Strategy, Practice, and Science* 1, no. 3: 155–59.

"The Unborn and the Born Again." 1977. *New Republic*, July 2, 5–6.

United Nations Development Program (UNDP). 1994. *Human Development Report 1994: New Dimensions of Human Security.* New York: Oxford University Press.

U.S. Congress. 2002. "Public Health Security and Bioterrorism Preparedness and Response Act of 2002." 107th Congress of the United States of America. Washington, D.C.: U.S. Congress.

U.S. Department of Energy. 2004. *Office of Science Strategic Plan, February 2004.* Washington, D.C.: U.S. Department of Energy.

Van der Westhuizen, Janis. 2005. "Arms over AIDS in South Africa: Why the Boys Had to Have Their Toys." *Alternatives* 30: 275–95.

Varela, Francisco J., and Antonio Coutinho. 1991. "Immunoknowledge: The Immune System as a Learning Process of Somatic Individuation." In *Doing Science: The Reality Club*. Edited by John Brockman, 239–56. New York: Prentice Hall Press.

Vernadsky, Vladimir I. [1929] 1998. *The Biosphere*. Reprint, New York: Springer.

Waldby, Catherine. 1996. *AIDS and the Body Politic: Biomedicine and Sexual Difference*. London: Routledge.

———. 2002. "Stem Cells, Tissue Cultures, and the Production of Biovalue." *Health: An Interdisciplinary Journal for the Social Study of Health, Illness, and Medicine* 6, no. 3: 305–23.

———. 2006. "Umbilical Cord Blood: From Social Gift to Venture Capital." *Biosocieties* 1: 55–70.

Waldby, Catherine, and Melinda Cooper. 2007. "The Biopolitics of Reproduction: Post-Fordist Biotechnology and Women's Clinical Labor." *Australian Feminist Studies* 22, no. 54.

Waldby, Catherine, and Robert Mitchell. 2006. *Tissue Economies: Blood, Organs, and Cell Lines in Late Capitalism*. Durham, N.C.: Duke University Press.

Waldby, Catherine, and Susan Squier. 2003. "Ontogeny, Ontology, and Phylogeny: Embryonic Life and Stem Cell Technologies." *Configurations* 11: 27–46.

Watanabe, Myrna E. 2001. "Can Bioremediation Bounce Back?" *Nature Biotechnology* 19: 1110–15.

Weber, Max. [1904–5] 2001. *The Protestant Ethic and the Spirit of Capitalism*. Translated by Talcott Parsons. New York: Routledge.

The White House (George W. Bush). 2001. "Remarks by the President on Stem Cell Research." August 9. Washington, D.C.

———. 2002. "National Sanctity of Human Life Day, 2002: A Proclamation." Washington, D.C.

———. 2003. "President Details Project BioShield." Press release, February 3. Washington, D.C. Available online at http://www.whitehouse.gov/news/releases/2003/02/20030203.html (accessed March 2006).

Wilkie, Dana. 2004. "Stealth Stipulation Shadows Stem Cell Research." *The Scientist* 18, no. 4 (March 1): 42.

Williams, Brian, Eleanor Gouws, Mark Lurie, and Jonathon Crush. 2002. *Spaces of Vulnerability: Migration and HIV/AIDS in South Africa*. Cape Town, South Africa: Idasa.

Wohlstetter, Roberta. 1962. *Pearl Harbor: Warning and Decision*. Stanford, Calif.: Stanford University Press.

Wood, Carl, and Alan O. Trounson. 1999. "Historical Perspectives of IVF." In *Handbook of In Vitro Fertilization*. Edited by Alan O. Trounson and David K. Gardner, 1–14. 2nd edition. London: CRC Press.

Woodger, J. H. 1945. "On Biological Transformations." In *Essays on Growth and Form Presented to D'Arcy Wentworth Thompson*. Edited by W. E. Le Gros Clark and P. B. Medawar, 94–120. Oxford: Clarendon Press.

Woodward, Bob. 2001. "CIA Told to Do 'Whatever Necessary' to Kill Bin Laden. Agency and Military Collaborating at 'Unprecedented' Level;

Cheney Says War against Terror 'May Never End,'" *Washington Post*, October 21.

World Health Organization (WHO). 2000. *WHO Report on Infectious Diseases: Overcoming Antimicrobial Resistance.* Available online at http://www.who.int/infectious-disease-report/2000/ (accessed March 2006).

World Trade Organization (WTO). 1996. *Trade-related Aspects of Intellectual Property (TRIPS) Agreement.* Geneva: WTO.

Wright, Susan. 1990. "Evolution of Biological Warfare Policy, 1945–1990." In *Preventing a Biological Arms Race.* Edited by Susan Wright, 26–68. Cambridge, Mass.: MIT Press.

———. 2004. "Taking Biodefense Too Far." *Bulletin of the Atomic Scientists* (November–December): 58–66.

Youde, Jeremy. 2005. "The Development of a Counter-Epistemic Community: AIDS, South Africa, and International Regimes." *International Relations* 19, no. 4: 421–39.

Zelizer, Viviana A. 1979. *Morals and Markets: The Development of Life Insurance in the U.S.* New York: Columbia University Press.

Zeller, Christian. 2005. "Innovationssysteme in einem finanzdominierten Akkumulationsregime—Befunde und Thesen." *Geographische zeitschrift* 91, nos. 3–4: 133–55.

Index

Achmat, Zackie, 57
acts of God. *See* catastrophe risk
Agamben, Giorgio, 63, 98, 192n7
Aglietta, Michel, 23, 192n3
AIDS. *See* HIV/AIDS
Ammerman, Nancy T., 191n2
Anderson, Philip W., 44
animation, 106–11; cinematography and, 107, 109–110; resurrection and, 125; suspended, 13, 109–10, 125, 127
anthrax, 74, 87, 94
Aquinas, Thomas, 158–59, 192n4
Aradau, Claudia, 183n14
archaea, 34
Aristotle, 5, 15
Arnott, Jayne, 183n24
Arrighi, Giovanni, 19, 29, 58, 179n3
Arrow, Kenneth J., 44
Arthur, Brian W., 44
Ashcroft, Frances, 34
Ashforth, Adam, 71, 183n19
assisted reproductive technologies (ARTs). *See* reproductive medicine
astrobiology, 21, 40–41, 181n26
Auger, Francois A., 187n6
automata, 106, 109
autopoiesis, 35, 38. *See also* capital; self-organization
Avant, Deborah D., 185n20
avian flu, 94

Bacher, Jamie M., 91
bacterial recombination, 33–35. *See also* horizontal gene transfer
Bak, Per, 38

Bakker, Isabella, 59, 135
Bauman, Zygmunt, 60
Bayh-Dole Act, 27
Bell, Daniel, 17, 176
Benjamin, Walter, 193n10
Bensaïd, Daniel, 179n2
Bergson, Henri, 181n22
Beveridge, William, 8
Bichat, Xavier, 5–6
Biggers, J. D., 134
Bigo, Didier, 182n13
Billingham, R. E., 125
biochemistry, 105
biodefense research: BioShield project and, 93, 186n25; incorporation into public health, 93, 97–98; in United States, 91–92, 97–98. *See also* biological warfare
bioeconomy, 18–21, 25, 45–48, 149, 175; as response to oil depletion, 47–50
biofuels, 41, 49
biogeochemistry, 35
Biological and Toxins Weapons Convention (BTWC), 84, 86, 184n8
biological security, 63–67, 70, 72, 80
biological sensors, 91
biological warfare, 93, 101; history of U.S. policy on, 84–88; South African program in, 68. *See also* biodefense research
biology: historical development of, 5–7, 31–34; mechanistic theory of, 104, 106–11, 115. *See also* animation; developmental biology; life sciences; microbiology; molecular biology
biomedical technology. *See* biotechnology industry

Breinigsville, PA USA
28 February 2011
256498BV00003B/1/P